Heritage in the Home

This book explores how people encounter the pasts of their homes, offering insights into the affective, emotional and embodied geographies of domestic heritage.

For many people, the intimacy of dwelling is tempered by levels of awareness that their home has been previously occupied by other people whose traces remain in the objects, décor, spaces, stories, memories and atmospheres they leave behind. This book frames home as a site of historical encounter, knowledge and imagination, exploring how different forms of domestic 'inheritance' – material, felt, imagined, known – inform or challenge people's homemaking practices and feelings of belonging, and how the meanings and experiences of domestic space and dwelling are shaped by residents' awareness of their home's history. The domestic home becomes an important site for heritage work, an intimate space of memories and histories – both our own but also not our own – a place of real and imagined encounters with a range of selves and others.

This book will be of interest to academics, students and professionals in the fields of heritage studies, cultural geography, contemporary archaeology, public history, museum studies, sociology and anthropology.

Caron Lipman is an honorary research fellow at Queen Mary University of London. A cultural geographer and heritage practitioner, her interests span contemporary engagements with the past, belonging, identity and home, and critical geographies of belief and folklore.

Critical Studies in Heritage, Emotion and Affect
In Memory of Professor Steve Watson (1958–2016)

Series Editors: Divya P. Tolia-Kelly (Sussex University) and Emma Waterton (Western Sydney University)

This book series, edited by Divya P. Tolia-Kelly and Emma Waterton, is dedicated to Professor Steve Watson. Steve was a pioneer in heritage studies and was inspirational in both our personal academic trajectories. We, as three editors of the series, started this journey together, but alas we lost his magnificent scholarship and valued counsel too soon.

The series brings together a variety of new approaches to heritage as a significant affective cultural experience. Collectively, the volumes in the series provide orientation and a voice for scholars who are making distinctive progress in a field that draws from a range of disciplines, including geography, history, cultural studies, archaeology, heritage studies, public history, tourism studies, sociology and anthropology – as evidenced in the disciplinary origins of contributors to current heritage debates. The series publishes a mix of speculative and research-informed monographs and edited collections that will shape the agenda for heritage research and debate. The series engages with the concept and practice of Heritage as co-constituted through emotion and affect. The series privileges the cultural politics of emotion and affect as key categories of heritage experience. These are the registers through which the authors in the series engage with theory, methods and innovations in scholarship in the sphere of heritage studies.

Museum Franchising in the Age of Cross-Border Heritage
Beyond Boundaries
Sarina Wakefield

Visualising Place, Memory and the Imagined
Sarah De Nardi

Heritage in the Home
Domestic Prehabitation and Inheritance
Caron Lipman

For more information about this series, please visit: www.routledge.com/Critical-Studies-in-Heritage-Emotion-and-Affect/book-series/CSHEA

Heritage in the Home
Domestic Prehabitation and Inheritance

Caron Lipman

Routledge
Taylor & Francis Group

LONDON AND NEW YORK

First published 2020
by Routledge
2 Park Square, Milton Park, Abingdon, Oxon OX14 4RN

and by Routledge
52 Vanderbilt Avenue, New York, NY 10017

Routledge is an imprint of the Taylor & Francis Group, an informa business

British Library Cataloguing-in-Publication Data
A catalogue record for this book is available from the British Library

Library of Congress Cataloging-in-Publication Data
Names: Lipman, Caron, author.
Title: Heritage in the home : domestic prehabitation
and inheritance / Caron Lipman.
Description: Abingdon, Oxon ; New York, NY : Routledge, 2020. |
Includes bibliographical references and index.
Subjects: LCSH: Dwellings–Social aspects–England–Case studies. |
Material culture–England–Case studies. | Memory–Sociological
aspects–Case studies. | Local history–Case studies. |
History–Methodology–Case studies. | Great Britain–Social life
and customs–Case studies. | Great Britain–Historiography–Case studies.
Classification: LCC GT285 .L57 2020 (print) | LCC GT285 (ebook) |
DDC 392.3/60942–dc23
LC record available at https://lccn.loc.gov/2020001312
LC ebook record available at https://lccn.loc.gov/2020001313

ISBN: 978-1-138-61608-0 (hbk)
ISBN: 978-0-429-46248-1 (ebk)

Typeset in Times New Roman
by Newgen Publishing UK

Contents

List of figures   vi
Acknowledgements   vii

1 Introduction: heritage in the home in context   1

PART I
Experiencing the past at home   43

2 Knowing and imagining the past at home   45

3 Presences of the past: energies, auras, ghosts   68

PART II
Past residents at home   89

4 Connecting with the past: domestic genealogies   91

5 Belonging to home: negotiating ownership   117

PART III
Material pasts at home   145

6 Found objects: the tangible past at home   147

7 Improving home: the ethics and aesthetics of custodianship   170

Conclusion: heritage in the home in wider context   198

Index   225

Figures

1.1 A carved wooden pig found under floorboards reflecting a
room's previous use by children   27
2.1 Outlines on walls show where pictures would have hung   65
3.1 A gargoyle-like green man figure on the corner wall of
Sandra's home   84
4.1 Ben described his sense of connecting with previous residents
when holding onto the original banister, triggering questions
about their lives   113
5.1 This collection of small found objects was placed into a plastic
tub over time   142
6.1 Fragment of a biblical quote ('the wages of sin is death, but
the gift of god is eternal life') found on a wall near the front
door of Pam's remote farmhouse; she believed this reflected
the dale's previous Quakerism. The fragment was rescued
during restoration work and is kept in a box   166
7.1 This original door handle in a living room became a focus
during one guided tour   194
8.1 The end of an old pipe found new use as a hook for a
dressing gown   219

All images © the author

Acknowledgements

Perhaps / The truth depends on a walk around a lake ... Perhaps there are times of inherent excellence / As when the cock crows on the left and all / Is well, incalculable balances ... not balances / That we achieve but balances that happen ...
Wallace Stevens, 'Notes Toward a Supreme Fiction'

I have many people to thank: firstly, all of the people I interviewed, who invited me into their homes, allowed me a glimpse into their lives, and entered into the spirit of the project so generously. I am excited to be part of this important Routledge series edited by Emma Waterton and Divya P. Tolia-Kelly, and I would like to thank them – and the anonymous reviewers – for their very helpful suggestions. My family and friends, as ever, for all their support, particularly my mother Frances, my wonderful (and patient) partner Sigrid, and our beautiful children, Siri and Tobi (for more unexpected delights!). I have benefitted from the generosity of a wide community of scholars – some I have known personally and others whose work has inspired me. The Geography department at QMUL offered me a stimulating home for many years; I thank Professor Alison Blunt and Professor Alastair Owens for helping to develop this, and other, projects. My most heartfelt thanks goes to my main collaborator, the brilliant Professor Catherine Nash, who mentored me through the various stages of postgraduate and postdoctoral research and the lectureships which followed. It was Catherine who thoughtfully nudged me to fulfil my potential. I hope others starting on a similar journey are as lucky. Lastly, during the process of researching this book, I co-curated an exhibition at the Geffrye Museum of the Home in east London and I'd like to thank the group of participants who agreed to work closely with me in bringing this to fruition, and the museum's gifted curators at that time, Eleanor John and Alex Goddard. Alex has been in my thoughts as I wrote this – another beautiful soul I was lucky enough to meet, albeit too briefly.

1 Introduction

Heritage in the home in context

For many people, the intimacy of domestic dwelling is mediated by varying degrees of awareness that their home has been the residence of prior occupants whose traces remain in its physical fabric and in the objects left behind. This book explores this near-ubiquitous but overlooked experience in the context of a range of English households, framing home as a site of historical encounter, knowledge and imagination. It offers insights into the affective and embodied experiences of heritage at the level of everyday homemaking, exploring how the meanings and experiences of domestic space and dwelling are shaped by residents' awareness of their home's history and how this history informs, challenges or enhances feelings of belonging. Domestic inheritance is defined in relation to the objects, aesthetics and arrangements of space encountered or discovered in the course of homemaking, and through the home's stories, memories and atmospheres.

The book thus positions the home as an important site for heritage work, an intimate space for the materialisation of memories and histories not our own, of imaginary encounters with selves and others, a space which demands that we engage critically with how heritage is lived, practised, felt and experienced and the meanings and values imbricated in emotional and embodied responses to the home's past. The book asks: How do inhabitants negotiate a sense of belonging and relatedness in the light of actual or imagined previous inhabitants, and the presence of the home's inherited objects, spaces or atmospheres? What are the relationships between memory, the historical imagination, affect and emotion in the material and embodied processes of 'making home'? What pasts engage residents, what values are granted to them and for what purpose? And how might a focus on the past at home extend our understanding of the ways heritage is theorised, imagined and practised?

The book builds upon arguments for the importance of attending to the affective, emotional and embodied registers of experiences of heritage and memory.[1] It draws upon 'understandings of the world' that are 'lived, embodied and tangled up with how we do things, our doings and our enactments in the moment' (Waterton 2014: 826). It attends to the ways subjects and objects are formed through their encounters with each other (Ahmed 2004b), drawing upon recent cultural theory on subjectivity as relational, performative and emerging through human and more-than-human others (Lorimer 2005).

Focusing on responses to the spaces of heritage in the context of the domestic interior enables a detailed and nuanced exploration of the meanings, values and implications of such registers of experience in relation to the active and ongoing process of making home. Such a focus also raises questions about the nature of 'everyday' and 'lived' forms of heritage practice and experience as well as more 'popular' forms of historical engagement – often positioned in opposition to institutional or 'expert' practices, processes of knowledge formation, scope and outputs. The book extends to the site of home current interest in imaginative, emotional and embodied encounters at heritage sites and within historical research. Unlike most institutional sites of heritage, the past at home took place *here*, within the intimate micro-spaces of the domestic interior, intrinsic to the continuous and performative process of securing senses of subjectivity, belonging and relatedness. The choices people make when attending to their home's inheritances impact on the act of making home itself; attending to these reveals particular ways in which people attempt to bridge or reinforce the 'gap' – the perceived differences – between *then* and *now*. In turn, this process cannot be separated from perceptions of the differences and similarities between *us* and *them*, even as these shape us. Thus we can consider the various experiences of and practices around the past at home for what they expose of forms of relatedness and emotional responses to past others – those from the 'deeper' past as well as more full-bodied encounters with recent and future residents. Of particular interest is the fact that previous residents, at least at the point of entering a new home, are usually *unknown* – strangers whose perceived differences or similarities to us are part of the negotiation required in making a pre-inhabited space our own. How we engage with those who are 'not us' – not held within our own memories, not narrated in our own histories – becomes part of a broader ethical project in which responses to past others expose values, beliefs and feelings which motivate acts of assimilation or exclusion.

In the following sections I will work through the ideas, debates and questions arising in relation to the three main themes of this book: (i) defining the parameters of 'everyday' heritage within the domestic interior; (ii) exploring the forms and effects of self-other relationships emerging from encounters with heritage at home; and (iii) how responses to such encounters are expressed in terms of knowing, imagining and feeling as well as in relation to more-than rational experiences and beliefs.

'Everyday' heritage in the home

Beyond deliberately designed or designated sites of 'heritage', *everyday* engagements with the past also need to attend to the social and political dynamics of, as Lorimer (2005: 125) puts it, 'how life takes shape and gains expression in shared experiences, everyday routines, fleeting encounters,

embodied movements, pre-cognitive triggers, practical skills, affective intensities, enduring urges, unexceptional interactions and sensuous dispositions'.
The 'everyday' is a rather loosely-defined word, described by Susie Scott
(2009: 3) as 'nebulous, pervasive and ambiguous: obvious to the point of
elusiveness'. The 'everyday' has become a signifier for particular forms of
experience, and in turn shorthand for political arguments in relation to: the
importance of attending to momentary 'lived' encounters; the quotidian,
mundane or ordinary 'life' (or the 'ordinary' people living it); and the geographically 'local', near or micro-scale (everyday or ordinary *spaces*).

Denis Byrne (2013: 596) has critiqued the 'privileging in heritage practice
of national scale over local scale', whilst Emma Waterton points to the way
'people interact – routinely and creatively – with heritage' which not only
'allows us to address critiques of heritage practice from gendered, non-elite,
vernacular and post-colonial perspectives' but also widens the scope of heritage enquiry to include both 'grander settings of large museum spaces and
in everyday life' (Waterton 2014: 829). David Crouch (2015: 188) adds that
'institutional sites' of heritage are 'merely fragments in a wider and deeper
mixture of prompts, memories, other sites, relations and feelings in our own
lives, as "ordinary" individuals going about our living ... We participate in
making our own heritage in acts and feelings of everyday living ... Heritage is
not detached' from the rest of our lives.

This focus emerges from Michel de Certeau's (1998, 2011) emphasis on the
practices of 'everyday life', whilst Pierre Nora's (1996) 'sites of memory' included
mundane places and Raphael Samuel ([1994] 2012) argued for engagement with
quotidian heritage in order to hear 'different voices', including a 'shift beyond
the museum' (Atkinson 2008: 374). The widening scope of participation and
collaboration in deciding what is meaningful of the past, proliferation of forms
of engagement in cultural memory, and attention to gender, race, sexuality and
other identities, has led to a greater focus upon what 'people and communities
do, with or without official sanction' (Atkinson 2008: 385; Smith 2004) and the
'mundane aspects of daily practice' (Atkinson 2008: 387), including the 'micro-
scale subjectivities of everyday life'. People are no longer just consumers of
heritage but are 'taking control of making and interpreting the past' (Ashton
2010: 11–12) to the point that 'resisting official narratives is [seen to be] acceptable and, indeed, productive and positive' (Atkinson 2008: 397).

For some, a focus on the 'everyday' becomes a counter-hegemonic move, a
form of 'dissonant' heritage. Iain Robertson argues that 'vernacular' heritage
or 'heritage from below' focuses on the 'role of non-elites' and the routine
habits of quotidian and mundane spaces (Robertson 2015; 2016 [2012]). He
argues that heritage literature is 'seduced' by the 'nationalist, top-down, commercial and tourism-focussed perspectives' (Robertson 2016 [2012]: 1), whilst
heritage work should actually be 'about people ... their sense of inheritance
from the past and the uses to which this sense of inheritance is put'. This
creates 'anti-hegemonic possibilities' for 'expressions of identity and ways of
life that run counter to the dominant', pointing to 'alternative constructions

of the past' (10) which do 'not seek to attract an audience' but draw on 'the ordinary and quotidian' (2). Further to this, investment in heritage, it is argued, should include 'ordinary spaces where we live'. This allows for 'more comprehensive understandings of the complex social memories that pattern our contemporary world' (Atkinson 2008: 392), given that the 'local context' is where the relationship 'between heritage and identity establishment and maintenance is often most meaningful' (Robertson (2016 [2012]: 6).

This focus is core for the subject of this book. But perhaps the quotidian has been given too much work to do as a counter to commercial or official forms of 'heritage'. Certainly, Robertson's description of a sense of inheritance that does 'not seek to attract an audience' is useful for considering the differences between institutional and everyday practices – in relation to the different degrees to which 'heritage' is designed to be performed for others to experience. But this does not preclude the 'past' being displayed or performed in other ways, or indeed a consciousness of 'others' in enacting responses to the 'everyday' past.

Secondly, conflating 'ordinary' and 'everyday' heritage with 'ordinary' or 'everyday' *people* might reinforce a binary between experts and non-experts (or 'amateurs'), heritage professionals and 'the people', whilst romanticising the latter's experience of the past as somehow more authentic. Bella Dicks (2000: 37), for example, describes how vernacular forms of heritage practice 'address the ordinary and the everyday in order to offer "ordinary people now" the chance to encounter "ordinary people then"'. There is a danger that people's experiences are appropriated as anti-hegemonic as if everyday encounters with the past should only be considered deliberate counters to powerful institutional 'others' by default of being 'everyday'. The risk in so doing is that this might pre-empt the values underpinning, and outcomes of, such encounters, and miss the complexity and variety of experiences of the past in different settings.

Lastly, assuming that a focus on the 'local' is inherently political and exists separately from other ways heritage might be scaled or scoped forces a divide between everyday and institutional practices which might rather be considered part of a continuum, a spectrum of engagements. Museum practitioners, for example, increasingly acknowledge the 'need to take account of the visitor experience inside *and outside* the walls of the museum' (Watson 2018: online; my italics). But, as Moussouri and Vomvyla (2015: 99) argue, there have been few studies looking at 'how people use the past in their everyday life or how they make sense and engage with the past – especially the more recent past – across different contexts'. They call for more research into 'how people use and make sense of the past, using their *own frames of reference*' (ibid; my italics) – the subject of this book.

Placing home

David Atkinson's example of 'mundane' heritage focuses on the perhaps rather public space of a *dockyard*, prompting Robertson (2015: 997) to correctly counter that this raises 'important questions over the nature of the mundane and its particular space as mundane heritage'. Within this orbit is the home, which Robertson in turn describes vaguely as 'relatively mundane', maintaining that 'houses, quotidian and vernacular' are important sites of memory work which heritage studies have 'failed to pay sufficient attention to'. Robertson is correct here, but the association of houses with the 'mundane' (or, indeed, the broader range of spaces which make up 'homes') implies something dull, uneventful, uninteresting, parochial. Homes, with their specific historically and culturally evolving forms, social complexity and range of emotional impacts, cannot, in this sense of the word, be considered mundane spaces.

Home is, firstly, a physical site in which people live, a site to express, constitute and perform the self through its material cultures (Harrison 2007; Cloke and Jones 2004; Jacobs and Smith 2008; Miller 2001), and simultaneously an 'idea and an imaginary that is imbued with feelings' – of 'belonging, desire, and intimacy' and also 'fear, violence and alienation' (Blunt and Dowling 2006: 2). The historical development of the 'private' Western interior has been mapped onto the emergence of a distinct Enlightenment 'self'. This culturally specific and traditional ideal of the Western home as a place of exclusive belonging, privacy, familial intimacy, comforting routines and ownership has been challenged in the face of people's diverse realities and examples of the 'unhomeliness' of home for many (Douglas 1991: 262; Chapman and Hockey 1999; Rubenstein 2001). As a social space, home is experienced differently depending on gender, class, race, sexuality, generational differences, and 'diverse experiences of previous and current homes and the movement between them' (Al-Ali and Koser 2002: 6).

Humanistic ideas stressing a distinctiveness between home's insides and outsides continue to be influential, despite reinforcing a binary of public and private space feminist scholars have worked to undo. Yi-Fu Tuan (1977: 107), for example, reflected on the 'emotional temperature' distinguishing between 'intimacy and exposure ... [and] private life and public space', whilst Edward Casey distinguished between 'thick' and 'thin' places, where the 'former are suffused with a sense of sensual, emotional, and affective belonging that is embedded over time through repetitive practical, embodied engagement' (Edensor 2012: 1106; Bachelard 1994).

In contrast, home has been theorised through ideas of place as mobile processes that 'extend well beyond the confines of a particular place' (Cresswell 2004: 50), part of 'open and porous networks of social relations' (Massey 1994: 121). This has become a point of contention, leading as it has to claims that homes as fixed places – such as reflected in having a 'name, a boundary, and distinctive social and physical qualities' – carry negative connotations, representing exclusivity and reactionary politics for some (Harvey 1996: 293). Elsewhere, others have challenged the figuring of contemporary experience as characterised through rootless mobility in contrast to fixity and rooted belonging (Nash 2008: 11), given the 'association of home and rootedness

with political conservatism and the equation of mobility with cultural dyna-
mism, as well as progressive politics, that has accompanied [the] conceptual
counterposing of stasis and movement'.

As this book will show, home is a site of negotiation between ideas of fluidity and sedimentation, continuity and change. This can incorporate more
fluid notions of home as a 'dynamic process, involving the acts of imagining,
creating, unmaking, changing, losing and moving "homes"' (Al-Ali and
Koser 2002: 6; Gregson *et al.* 2007; Kaika 2004). Indeed, research on migrant
experiences has situated home as multi-local and trans-local, a 'space in which
the complex, multi-directional and often contradictory manifestations of
transnationalism are played out' (Sheringham 2010: 64). And home is also
more broadly defined in relation to the 'ability to feel at home in the city ...
shaped by migrations and other mobilities', with 'urban dwelling and mobility
... intertwined rather than separate' (Blunt and Sheringham 2018: 1).

These conceptions of home do not preclude continuing interest in the
domestic *interior* in relation, for example, to a scrutiny of the conflation of
'home, household and family' in masking tensions around the role of women
within domesticity and the 'complex constraints of privacy and choice'
(Hurdley 2013: 14), or questioning assumptions that homes are an uncomplicated place of belonging rather than critically engaging with the ways they
might be 'made to shelter some bodies and not others' (Schmitz and Ahmed
2014: 101). For some, home continues to be a potential site of 'dignity and
resistance' for women (Young 2005:146; hooks 1990; Pratt and Hanson 1994),
including of 'refuge from the alienating experience of racism and marginality'
for black women (Tolia-Kelly 2004: 316–317).

Homemaking practices

Exploring the domestic interior also focuses attention on its physical and
embodied aspects, including its material spaces, objects and homemaking
practices. Firstly, people undertaking renovation work might have, as Thomas
Yarrow has observed, a particular experience of their home's past 'corpor-
eally and viscerally ... the physical intimacy of working in these buildings
engenders a specific way of understanding and caring for their past' (Yarrow,
2019: 9). Dawn Lyon (2013: 29) has observed builders at work undoing the
layers of a building by removing 'other people's labour'. She noted how they
'live and breathe it, quite literally, as its dust and paint and debris get under
their nails and skin, and into their hair and eyes ...Their understanding of the
space emerged through working in and on it, coming to know it in this way,
over time' – a form of 'tacit knowledge' (35).

In contrast, much literature on heritage related to the home tends to understand buildings narrowly as 'material embodiments of specific pasts' (Yarrow
2019: 3–4), with practices focused on conservation 'as a way of protecting
these inherent or authentic qualities'. These conservation practices, however,
cannot explain 'how and why the past matters', which requires more than just

recognising the 'qualities that already exist' but also viewing such practices as 'social and discursive', involving acts of 'constructing those qualities and of choosing to value them in specific ways' (4). Such processes also involve an 'everyday negotiation of ethics', including questions such as whether the 'character of a building resides only as an embodiment of past people and events' or can be 'enhanced as it is modified in response to the lives of contemporary inhabitants' (6). Granting the home its own moral agency also leads, Yarrow adds, to a sense of responsibility to it (13). This is a useful framework for exploring what values, attitudes and ideas people bring to their own, broader, homemaking practices.

In a related analysis – of residents living in a 'historic district' in Arizona, North America – Jennifer Kitson and Kevin McHugh (2015: 488) argue that encounters with the past embody sensory domestic practices through which 'ideas, memories and imaginings' emerge based on 'bodily attentiveness to what feels right'. These dictate decisions about what to repair, restore, maintain and furnish more than any concern for 'architectural and historical accuracy' or 'adherence to strict preservation guidelines or historical fact' (488). The focus on the past of these homes was, they argued, a nostalgic form of 'everyday enchantment' (494) which allowed people to engage in 'micro-movements' from 'tip-toeing around creaky floorboards, polishing silver, carefully cleaning uneven surfaces, to wondering how a particular architectural oddity came to be'.

These examples are useful for reflecting on the different values involved in conservation practices and the role of 'bodily attentiveness' in making choices in dealing with the past's materialities at home. But they focus exclusively on middle-class home owners, raising further questions about responses to the past for those who do not own their homes or don't have the money to undertake conservation projects, as well as those who bring different cultural needs, concerns or values to the process of homemaking (Lawrence-Zuniga 2016; Yarrow 2019). What is the meaning given to 'conserving' the past for these people? Issues of class and race also need to be taken into account when considering preservation practices. Denise Lawrence-Zuniga, for example, shows (again in an example from North America) how 'predominantly white, middle-class' people favoured an 'idealized "traditional" image' (2014: 819) when restoring houses to their original condition, with 'alternative expressions of home' being suppressed in order to 'reproduce the re-imagined white suburb'. Home remodelling, she adds, can 'render intangible phenomena like historic values and significance concrete and durable' (Lawrence-Zuniga 2016: 7). Attending to people's homemaking practices is important for understanding what they value of the past and why, but we also need to take into consideration the wider contexts in which these are undertaken.

Memory and home

Memory is also seen as central in shaping imaginative and material geographies of home, with home 'profoundly linked with imagination and remembering'

(Findlay 2009: 116), particularly 'from the perspective of exile' (198). In turn, heritage is also associated with memory – both indeed considered 'partial, subjective, contested, political, subject to particular historical contexts and conditions, and thus dynamically changing' (Sather-Wagstaff 2011: 191; Sather-Wagstaff 2015; Hanlon, Hostetter and Post 2011; Radstone and Hodgkin 2003; May and Thrift 2001; Studdert and Walkerdine 2016).

Previous work tended to emphasise home as a realm of private memory, closely linked to senses of attachment. Bachelard (1994 [1958]: 6), for example, described how the 'memories of the outside world will never have the same tonality as those of home and ... our memories of former dwelling-places ... remain in us for all time'. For Michel de Certeau, likewise, the memories of 'successive living spaces never disappear completely; we leave them without leaving them because they live in turn, invisible and present, in our memories and in our dreams' (1998: 148; see also Halbwachs 1994 [1925]). Memory here is an important aspect of the construction of private senses of belonging – feelings assumed to be 'anchored in particular places and territories such as neighbourhoods or nations in which people feel "at home"' (May 2017: 401).

Others insist that memories require 'sustained social, interpersonal interaction in order to endure' (Sather-Wagstaff 2011: 191). Joy Sather-Wagstaff also directs us to consider how memories manifest performatively within specific domestic and everyday spaces, a form of 'memorywork', the 'enactment of daily routines' within 'everyday, lived social life'. People, Robertson (2016 [2012]: 17) adds, 'articulate and construct their sense of the past and historical identities', through 'performed' routines within domestic space. This continuing influence of Bergsonian ideas of habitually-practiced embodied memory – 'habits, traditions and rituals ... sedimented in the body' (Dragojlovic 2015: 321; Connerton 1989; Waterton 2014) – suggests senses of belonging and identity to place emerge through continuous re-enactments (Seamon 1980), places themselves being sedimented through 'habit memories' (Young 2005: 139–149; Wetherell 2012), and affects 'conditioned by previous experience, by habit, by familiar emotions and sensations that produce feelings of belongingness or otherwise' (Edensor 2012: 1114; see Bourdieu 1977).

Others have explored broader contexts, including the 'significance of memory, identity and emotion' in shaping everyday domestic and urban life in diasporas and transnational spaces (Blunt and Bonnerjee 2013: 220–221), and the continuing impacts of colonialism and slavery in relation to the ways in which memories and emotions circulate and take form within the home (Jesser 1999; Nasta 2002). Linear senses of homes as temporally successive are also disrupted in the transnational experiences of migrants, 'complicating distinctions between home as a structure of current residence and that of place of origin' (Sather-Wagstaff 2011: 198). Divya Tolia-Kelly's idea of 're-memory' (2004: 1) in her exploration of British Asian women's lives, describes how visual and material domestic culture is imbued with memory which signifies broader relationships between cultural identity, heritage and race. Re-memory 'is not an individual linear, biographical narrative', but a 'conceptualisation of encounters with memories' which are 'signifiers of "other" narrations of the past not directly experienced but which incorporate ... other's oral histories or social histories that are part of the diasporic community's re-memories'.

Material cultures of home

Research has also explored how historical activity 'imprints itself and leaves material traces' (Radstone and Hodgkin 2003: 8), and how memories are 'created from the material strategies we use' (Buchli and Lucas 2001: 79; Edensor 2001). Blakely and Moles (2019: 623) describe 'local, situated practices of memory work as a component of collective memory' whereby material traces of the past circulate through different spheres – from private, domestic realm to 'public' (see Blunt and Dowling 2006; DeSilvey 2007). A recent interest in the social conditions and effects of historical artefacts at home has also led to interest in the making and meaning of the past away from the public sphere (Buchli and Lucas 2001; Owen *et al.* 2010; Jeffries *et al.* 2009), whilst a focus on the material cultures of home has also offered insights into the role of *objects* in securing senses of belonging – one aspect of a range of interventions on the way objects 'generate social effects not just in their preservation and persistence, but in their destruction and disposal' (DeSilvey 2006: 324), with recent interest in the consumption and recycling of second hand and waste objects, including those things 'freed from the drudgery of being useful' (Benjamin 2002 [1935]: 18; Grossman 2015; Digby 2006). In 'binning, giving away, passing-on and selling the surplus, the excess, people are continually attempting to work out what to do with particular things, drawing on specific meaning frameworks and their conjunctures with the particularities of certain objects' materialities as they do so' (Gregson *et al.* 2007: 197). Through such practices of divestment, we 'continually re/constitute social orders, using what we do with and to things ... to constitute narratives of us, of others and our relations to them' (198).[2] But what of those objects which are not chosen, but are just *there*, waiting for us, drawing attention to our homes as previously inhabited by others? How do people's responses to *these* objects impact on the processes and practices involved in making home? How people deal with past objects they inadvertently inherit by moving into a particular building may create – or impede – senses of belonging at the heart of homemaking.

The home as archive

The home itself is not just a depository of the material past, but one with its own spaces and layers, as reflected in the double meaning of the archaic German word *heimlich*, which means 'homely' (in opposition to *unheimlich*, the uncanny or unhomely), but which – just like *unheimlich* things – also denotes something concealed, out of sight, *hidden* or secret (Freud 2003

[1919]).[3] In this sense home becomes a place of hidden things, or a place to hide things – to store as much as to display – a form of archive. Tim Cresswell (2012: 164) argues that 'archives are leaky places produced through a contested set of valuations concerning which objects count as worthy and significant'. Certainly it is accepted by many that within institutional settings, how (and what) objects are chosen to be displayed or stored reveals the values given to them. Jude Hill, for example, describes how the display of amulets in glass cases exhibited and archived at the Wellcome Collection was designed to subdue their 'latent' magic powers by categorising them as objects of archaic, pre-modern value within a 'narration of a broader scientific progress' (Hill 2007: 75). And yet, she argued, the museum 'could not expunge' their magic, and their containment in archives merely seemed to increase their power rather than remove it. This example draws attention to the different motives as well as impacts of displaying or 'storing' objects. But Cresswell (2012: 175) also calls for a 'broader definition of archives', to consider 'other kinds of collecting and other kinds of space as archival, including places themselves'. One such archive away from the institution includes the 'family archive' (Woodham *et al* 2017: 203). This is made up of 'documents, photographs, heirlooms, scrapbooks, recipes and a whole range of other items that reveal insights into past generations, and preserve family stories'. Such 'assemblages' may not be consciously thought of as 'archives' at all but constitute, it is argued, an 'important and undervalued site of meaning and identity construction' – an unofficial or 'behind the scenes' form of heritage. We might ask, what different forms of 'domestic archive' might be created with the objects people *find*? What is the range and variety of traces of previous habitation – overt and subtle, surface and hidden, fixed and mobile – encountered in the home? What impacts do they have? How do people deal with hidden or lost things that are then revealed? Are they stored, displayed, moved, thrown away? And what contingent factors influence the choices people make in dealing with them? How far do people become curators of the 'domestic archives' of their own homes – ones, nonetheless, made up of *other people's* left-over things rather than their own?

Relating to previous inhabitants at home

'To dwell means to leave traces', Walter Benjamin (2002 [1935]: 18) wrote: 'The traces of the inhabitant are imprinted in the interior. Enter the detective story, which pursues these traces'. Benjamin reminds us that the material traces left behind by previous inhabitants offer insights into past social contexts. These include a home's design, indicating 'broader social attitudes toward intimate places' (Davidson 2009: 334), and also the ways attitudes change over time, given the 'practices that gave rise to the size, form, and aesthetic of these homes are not – and cannot be – the same as contemporary practices and performances in maintaining them' (Kitson and McHugh 2015: 488).

Nonetheless, what has been less explored is the range of responses to past residents, beyond (if also imbricated within) the different layouts, material-ities and domestic practices revealed of the home's pasts. What emotional and other responses result from processes of researching past residents or otherwise knowing or imagining them? This book starts with the premise that such responses to these 'presences' of the past will have impact because the home is bound up in the identities of its inhabitants that it shapes. Rather than a private space, it is one full of imagined 'others' from the deep past or real 'others' who knock on the door and, as will be explored, leave their post, plants and objects behind. It is through such encounters that the past is imagined and reshaped for the present and future. Thus in order to examine the past at home, we need to attend to the bodily, emotional and affective responses to both the tangible aspects of the home's various rooms and spaces, the inherited décor and aspects of its physical fabric – and also its *less* tangible elements – the feelings, memories, atmospheres, ineffable experiences and imaginative responses to past residents themselves.

Commentary on the way people consider past lives at heritage sites suggests a 'paradoxical' desire to both experience them as different and to assume similarities (Hamilton and Ashton 2003: 5–30). But in relation to encounters with the past *at home,* there is a common assumption that reflecting on past residents will lead to one of two responses: a sense of enrichment, fascination and connection or, in contrast, feelings of intrusion, alienation and anxiety. The latter starts with the premise that, when moving into a new home, one needs first to cast out past others by creating home anew; home needs to be *made* homely and a place of uncomplicated belonging through a process of 'familiarisation':

> Though I may legally possess this abode, it does not belong to me at first. Somewhat like being a foreign face, the presence of the former occupant lingers about the house. If I am sensitive to this atmosphere, I am acutely aware of the alien texture of this reality. This new enclosure – the walls, the ceiling, the floor – is hard, crisp and foreign to the touch. It seems to face someone else, resisting me, failing to reveal itself. The labor of my caretaking turns this place into a home, into a place that addresses me as familiar, as belonging to me.
>
> (Lang 1985: 202)

The intimate encounter with the materiality of the space – 'hard, crisp and foreign to the touch' – is expressed here as something rough and harsh, symbolising a rejection, a pushing back.

The sense of unfamiliarity and alienation when moving into a pre-inhabited home can lead to an anxiety that the process of making home for oneself may not be fully possible. This is reflected in Daniel Miller's rather dramatic assumption of a problematic 'discrepancy between the longevity of homes and the relative transience of their occupants' with the attendant 'feelings of alienation' arising 'between the occupants and both their homes and their possessions' with which residents need to 'come to terms' (Miller 2001: 107). This assumption relates to a belief in the way objects at home are considered personal anchors of self-coherence, and a 'scaffold for memory' (Spelman 2008: 140); the presence of other people's objects thus becomes unnerving. In popular haunted house narratives, this generally becomes something menacing.[4]

Elsewhere, however, a rather different assumption is made: that getting to know one's home's past history has positive, benign, effects. In part, this reflects a belief in the importance of giving voice to history's forgotten or overlooked, through researching past lives at home. Elisabeth Skinner, for example, concludes that 'history could be about ordinary people' (2016: 110) in her reflection on a project to identify a village's local heritage, a 'search for the lives of ordinary people who lived in our houses and worked in our community. Until now, many of those people were the "disappeared" of history' (ibid: 111). Fenella Cannell (2011: 477) also challenges the belief that popular genealogy is solipsistic, arguing that an understanding of 'ordinary ancestors' is a form of 'care for the dead' (462) which includes a 'democratizing impulse ... the ownership and class positioning of knowledge' (475). By extension, exploring the ordinary people who once lived in your home might also be motivated by a desire for forms of knowledge which democratise the past.

In turn, popular biographies of particular homes can also tend to play up the similarities between 'ordinary' past and present residents, universalising their 'passions and griefs', as in this description:

> The vast mass of men and women in every time do not leave behind them either renown or testimony. These people walked our streets, prayed in our churches, drank in our inns ... built and lived in the houses where we have our being today, open our front doors, looked out of our windows, called to each other down our staircases. They were moved by essentially the same passions and griefs that we are. ... Yet almost all of them have passed away from human memory and are still passing away, generation after generation.
>
> (Tindall 2006: 6)[5]

In turn, popular house history research is also presented as an uncomplicated way of connecting to one's home. The popularity in the United Kingdom of BBC television programmes on this theme – including the series *A House Through Time*, 2018–, *The House Detectives*, 1997–2002, and the previous *Where Do You Think You Live?* strand of the ongoing BBC genealogy series, *Who Do You Think You Are?* – suggest this has become a dimension of knowledge, practice and meaning that deserves academic attention. Published guides helping 'amateur' historians explore the material and social histories of their homes also describe this practice in terms of a hobby for enthusiasts, often focusing on research as a 'detective trail' (Brooks 2007; see also Backe-Hansen 2011). The house is given agency as both special – 'almost every structure which has been in existence for fifty years or more is in some way special and has its own unique story to relate' (Bushell 1989: 10) – or requiring coaxing – 'rare is the house that gives away its history without a struggle' (Austin *et al.* 1997: 11). But it is the tracing of the home's 'family tree' (Bushell 1989: 64–72) which gives it 'personality and meaning' (ibid: 9): 'Your home has a lineage and during its history it has been home to any number of families. You may own your home, but perhaps it would be more accurate to view yourself as having a time share in history' (Austin *et al.* 1997: 9). In these guides, past residents are situated as part of an extended 'family' to which one can forge positive connections. The sense of shared belonging of inhabitants across the life of the house as part of a temporal sharing of ownership allows the researcher to 'empathize with the people who considered your house as their home as well' (Barratt 2001: 127). Once research is complete, the promise is that you will feel 'amazed at how the way you view your house has changed. It will increase your understanding of your home and you will love it more than ever' (Austin *et al.* 1997: 8). Only in one account is there a brief warning that the research may be at times unsettling: 'You may not always like the tales that you are told about your present home' (Barratt 2001:4).

Domestic genealogies

This book does not assume that an awareness of or knowledge about past residents is only ever rendered as either alienating *or* enhancing. Indeed, might people express a range of feelings, even simultaneously? One way to avoid limiting an examination of responses within the home is to focus on how present occupants *relate* in different ways to those who preceded them.[6] In order to fully explore the variety of responses to past inhabitants, then, this book starts with the premise that these are complexly inflected through practices of making relations. This approach extends Catherine Nash's exploration of the genealogical imaginaries and idioms mobilised when people define themselves relationally (Nash 2005, 2002), drawing attention to how ideas of familial relatedness, family trees and lineage are created (the lineage here a site-specific sequence of *residence*). Just as 'in popular genealogy what is inherited, owned and shared also includes family stories, memories, cultural traditions and genealogical information' (Nash 2008: 19), the home links strangers over time through the shared spaces of the domestic interior. This framework – which elsewhere we have called 'domestic genealogies' (Lipman and Nash 2019) – opens up questions of what is included or excluded when facing the challenge of experiencing imaginative co-presence and co-habitation in domestic space.

The domestic genealogy approach draws attention to the different temporalities of past habitation, including the more recent as well as the more distant pasts when relating to former occupants who are still alive or long dead, whilst it also offers potential to examine people's responses to questions of social, cultural, economic and material differences between themselves and former residents. This focus draws attention to how kinship responses are shaped by what Marilyn Strathern (1996: 518) has described as a 'Euro-American' culture of giving positive value to connections, including social connections. Again, the figuring of *house* histories – as exemplified in the above discussion of popular practices of house research – is similarly based on a broadly shared value system in which having a *connection* to someone – in this case, a former resident through a domestic lineage – has intrinsic value. Thus a 'connection' implies some shared basis for affinity whether that is imagined genealogically or, in this book's focus, through shared experience of living in the same space. But it also has to be *made* to matter. In other words, the assumption that connections are made with past residents requires further scrutiny. Who decides to connect with whom, under what conditions and to what ends?

Selves and others at home

The focus on emotional responses to similarities and differences between past and current residents also returns us to the ramifications of considering home a private space. This belief requires an 'attempt to expel alterity' and 'all forms of otherness, as inherently threatening to its own internally coherent self-identity' (Morley 2000: 6). Private spaces thus are seen as necessary to construct 'personal boundaries', protecting against the 'unsolicited emotional intrusions of the other', whilst also 'projecting outside of ourselves unwanted affects … in a process commonly known as "othering"' (Brennan 2004: 24). But if, as described, an understanding of past lives in the home reminds people that it is a 'shared' space, how possible, or desirable, is it to 'banish' these past others in order to secure a sense of subjectivity?[7]

My previous research on experiences of the domestic 'uncanny' suggested home as a 'multitude of different agents within', an 'accumulation of materialities and temporalities', where 'the ghosts, the house's history, and the experiences and memories of current inhabitants … have to be accommodated because they collectively make the home what it is' (Lipman 2014: 85; Whatmore 1999; Haraway 1997; Hitchings 2004). Sara Ahmed's idea of the self as moulded by encounters with others was central for exploring these assumed accumulated memories and shared senses of belonging within the haunted home, given experience itself is framed as embodied and 'mediated by our continual interactions with other human and non-human bodies' (Ahmed and Stacey 2001: 5). Thus the home becomes both a 'social space' and a 'bodily space' (Ahmed 2000: 9) through 'strange encounters' within it – through the 'proximity of strangers who can no longer be seen as "outsiders"'. The domestic uncanny project explored how such 'others' were 'domesticated', recreated in the (often gendered) image of the self, reinforcing the strength of Ahmed's insight that 'in the gesture of recognising the one that we do not know, the one that is different from "us", we flesh out the beyond, and give it a face and form' (Ahmed 2000: 3).[8]

In the broader context of the different kinds of 'presences' of the past explored in this book, Ahmed's idea of 'affective economies' (2004a: 8) is particularly useful, drawing attention to how emotions are felt at an experiential level, attach themselves to and shape bodies, cultural practices and relations of power, including the way that 'different bodies, differently imagined, will have certain affective responses already mapped onto them, defined by social expectations and structures of feeling that have built up around issues of gender, class, race and so forth' (Waterson 2014: 829).[9] These ideas have formed the basis for exploring affective relationships of power at situated spaces of heritage sites, attending to how heritage itself is an intersubjective, 'complex and embodied process of meaning- and sense-making' (Waterton 2014: 824; Ladino 2015), and where heritage sites are considered as 'agents or co-participants/producers of a heritage experience'. Understanding the work that 'heritage sites, places and experiences *do* in wider social and political life', in terms of 'producing feelings of belonging, identity, inclusion, and by corollary, marginalization, subjugation and exclusion' becomes crucial for thinking about '*how* affective politics are reproduced in the context of debates about multiculturalism or social exclusion, for example, or *how* the conventional boundaries between us/them, self/other ought to be reconceived' (Waterton 2014: 831; Crang and Tolia-Kelly 2010).[10]

Within the home as a space which is shared rather than private – containing 'many different jostling actors' (Hitchings 2004: 169) – people might 'try on and play out roles and relationships of both belonging and foreignness' (Bammer 1992: vii). But, as Ann Varley (2008: 56) usefully argues, the self can 'relate to an other without assimilating or excluding it'. She cautions us not to become 'trapped in the binary of exclusionary or idealised space' when exploring the social contexts of home; housing the self 'need not always mean evicting the other' (58). Does an engagement with the past of one's home enhance or trouble senses of self and belonging? Does it draw attention to 'affective economies' of heritage in the home?

The home's memories not our own

The self/other encounter in temporal as well as spatial context requires us to unpack the assumed differences and similarities between *now* and *then* as a means by which to understand those between *us* and *them*. We might attend to how these encounters continue to be perceived as 'gaps' that need to be filled through emotion, imagination and knowledge. Whereas uncanny experiences became a form of 'surreal memory' (Lipman 2014: 66) – expressing something radically elusive, hesitant, unknowable, but nonetheless *experienced*, having palpable *effect* – a key consideration for this project is how subjectivity might be shaped by the memories and experiences of unknown others or events one has not experienced at first hand – those that are not *our own*.

Contemporary memory studies has its roots in theorising particular contexts of affective inheritance – how responses to past events might be 'passed on' to future generations of a family, with a particular focus on psychoanalytical ideas concerning how trauma is repressed but not fully forgotten (Hoskins 2003: 15). Abraham and Torok's (1994) theory of transgenerational haunting explored the 'gap in the conscious knowledge about family history that begins to haunt descendants' (Dragojlovic 2015: 320), whilst Marianne Hirsch's (2012) idea of 'postmemory' described the 'embodiment of pathology in individuals through the unconscious absorption of family narratives and ... the unspoken transmission of the parents' unconscious to the child' (Brogan 1998: 19). These ideas point to the possibility of how 'what seems unintentional is situated within the long history of violence, trauma and structural racism', and how people 'carry on in their bodies ... the often ambiguous, yet persistent presence of the past' – which nonetheless (in one view) they have the 'capacity to intervene into and disentangle from' (Dragojlovic 2015: 322).

These theories reflect how certain memories of events can get 'transmitted' across time, suggesting the possibility that, in a similar way, memories can pass down and across generations of shared residence, although not necessarily in response to traumatic events. The home's own 'memories' are both diffuse and palpable – inherited décor and found objects, atmospheres, smells – but nonetheless they may have power to make a deep impression which affects senses of self and belonging. Tonya Davidson (2009: 334), for one, describes this as a sense of the 'material hauntings' of houses through their inherited objects, which become the 'lingering residues of past owners'. But in her scheme, it is the *house* which is the agent of memory, not individual residents: 'houses embody memories beyond the lives of the current inhabitants'; they 'remember and haunt as they animate the memories of previous inhabitants, memories that become embodied by the houses and the current dwellers'.

Davidson assumes that encounters with *others'* pasts within the home *will* have an impact, but it is unclear what this impact might be. How might this complicate the continuing assumption about home and dwelling as an 'intimate hollow we have carved out of the anonymous, the alien', a 'movement from the strange to the familiar' (Lang 1985: 202)? How does it challenge the ongoing process of creating and maintaining subjectivity through senses of belonging, or disrupt embodied routines? How is home *made* familiar? (Lipman 2014; Blanco and Peeren 2010). Davidson utilises Alison Landsberg's theory of 'prosthetic memories', which also examines how memories which are *not our own* might get transmitted to us. These memories and 'the identities that those memories sustain' are not all 'shaped by lived social context' but rather a person can take on a 'more personal, deeply felt memory of a past event through which he or she did not live' (Landsberg 2004: 3). Landsberg believes this form of connecting to others can have a positive moral impact. Drawing on the work of post-Holocaust theorist Emmanuel Levinas, she argues that this creates 'the conditions for ethical thinking' by 'fostering empathy' (149); prosthetic memories allow people to 'feel connected' whilst 'recognizing the alterity of the "other"'.[11]

Landsberg does not focus on familial or private encounters but on the 'mass cultural technologies of memory' embodied in public cultural spaces such as museums and movie theatres. In fact, she is at pains to stress her interest is in the production of *public* memories in *public* spaces rather than a 'privatized past ... made meaningful to them because they, their family, or their ethnic group was in some measure part of it'. This implies a simplified separation of 'private' and 'public' in encounters between selves and others, the private sphere here synonymous with narrow tribal interests. However, her theory usefully draws attention to important questions: How far is it possible to experience others' feelings? Do people always develop senses of empathy, care and responsibility towards others through the act of recognising difference?

It is also useful to reflect on the aftermath of Landsberg's theory. Some have dismissed it as merely a metaphor for 'more conventional processes of cultural transmission – acquiring knowledge ... or responding with empathy to a narrative' (Berger 2007: 598). Others, drawing on Levinas and Derrida, point out that recognition of others is only ever *partial*, requiring acceptance of their 'independence and unknowability' (Varley 2008: 56) in order to develop a non-colonising relationship to the other. Landsberg's idealised 'empathetic' subject might end up 'suturing over an irreducible gap' that should be kept open in order to 'induce a moment of hesitation that would slow down Odysseus's imperializing speedboat' (Abel 2006: 387–8; see Shuman 2005). Elsewhere, Denis Byrne (2013: 596) accepts the 'potential to foster empathy with the experience of past others', but argues that this requires a 'sophisticated understanding of how objects become imbued with affect and how they transmit it', whilst Roger Simon (2005: 4–5) also suggests that public history requires a 'responsibility to the alterity of the historical experience of others – an alterity that disrupts the presumptions of the "self-same" ... an attentiveness to an otherness that resists being reduced to a version of one's own stories'. For Simon, however, it is important to recognise encounters with others and that to be 'touched by the past' is 'neither a metaphor for simply being emotionally moved by another's story nor a traumatic repetition of the past reproduced and re-experienced as present' (10).

Whilst this book does not assume that people's responses to the past lives of residents will always be empathetic, it is attuned to how encountering the past at home might lead participants to behaviours which reinforce 'otherness' or seek to overcome this, and to examine the ethical frameworks they themselves might bring to such a project. If Levinas impels us to acknowledge and accept the 'gap' between *us* and *them*, how far is this accentuated by the temporal distances between *now* and *then*? The book explores how people attempt to bridge or maintain this 'gap' in relation to encounters with both the earlier and more recent pasts and imagined futures.

Historical knowledge, imagination, emotion and the more-than rational

Dealing with the gap between *now* and *then*, in turn, requires that we understand how people define and value 'history', or the different 'pasts' which are deemed to make it up. North American newspaper *The Onion* (1997) once offered a cleverly satirical take on what was then considered an emerging trend – the craze for 'retro' objects and experiences. In a mock news story entitled, *U.S. Dept. of Retro Warns: 'We May Be Running Out of Past'*, it described a retro kitsch 'crisis', explaining how retro had 'always been separated from the present age by a large buffer of intermediate history' but recently the retro 'parabolic curve' had 'soared exponentially', and the gap was now 'rapidly shrinking'. Society had started expressing 'nostalgia for the decade in which we live' and was almost at the point of 'nostalgia for events which have yet to occur'. Retro trends, it concluded, 'threaten to consume the nation's past reserves faster than new past can be created'.

Part of the joke here was the playing with an assumption that the past is always behind us – a simple, linear backwards glance, vaguely 'before-now'. But are certain 'pasts' granted more significance than others? What influences the way different pasts are valued? And if 'the past' *is* always behind us, how do people deal with working out what can and cannot be known about it? What processes of getting to know (or not know) are people engaged in? And how do they define and value 'knowledge' itself? What is, for example, the place for emotional, sensual and imaginative experience in getting to 'know' the past? Accepting as a starting point that history is a 'culturally constructed form of knowledge with a heavily mediated relationship' to the past (Harvey 2015: 539), this book's focus on domestic encounters with the past offers one vantage point for critically reflecting on which forms of engagement with historical 'knowledge' are or should be granted meaning.

Public and private histories

Current interest in acknowledging and engaging with the 'agency of non-elite categories of people' (Harvey 2015: 917; Robertson 2016 [2012]; Ashton and Kean 2009; Smith, Shackel and Campbell 2011) has its origins in debates during the 1980s in the United Kingdom and Australia in which public historians challenged academic historians and academic historians, in turn, attacked popular heritage practices (Lowenthal 2015 [1985]; Wright 2009 [1985]; Hewison 1987; Samuel 1994; Ashton and Kean 2009; Smith 2004). Raphael Samuel turned to public history for a more democratic and inclusive engagement with history, complaining that 'knowledge had become locked up in professional academic circuits and thus became the preserve of the oligarchic few' (Robertson 2016 [2012]: 5). Focusing instead on the 'importance of local historians and the diverse, non-traditional range of materials they drew upon to construct their pasts' (Ashton 2010: 7), and celebrating 'unofficial knowledge' and 'popular memory', this work was enhanced by a North American survey of popular interest in the past (Rosenzweig and Thelen 1998) which 'demonstrated the complex ways in which people used the past in making themselves and their lives' (Ashton 2010: 3).

But the emergence of 'public history' generated debate in the United Kingdom. As some historians have defensively argued, Samuel's belief in history as 'the work of many hands' should *not* result in making 'critical academics redundant' (Strangleman 2013: 35). These historians have countered with a focus on the 'slippery' definitions of 'public', 'private' and 'personal' history – the latter here generally shorthand for historical engagement of little *significance*. Liddington (2002: 89–90), for example, suggested that:

> some public historians are surely just 'private historians' in cunning disguise. ... May 'private history' not include genealogists, some family or local historians, whose work not only starts from a personal interest but emerges as just that – the private history of a member of the public, still with little awareness of the needs of wider audiences or context?

Compared to this, she added, the 'practice of history is a discipline with the academic conventions of critical argument, evidence and citation' (91). In response, Paul Ashton (2010: 8) countered that public history 'involved the positive entanglement, rather than the separation, of the personal and the public' and that the boundary between them becomes 'increasingly permeable'. This brought into question, he said, 'if not undermined, the role of the academic historian' (Ashton 2010: 8): 'if we value some historical practices over others, it is because of historical decisions' (11–12).

These tensions inherent in deciding the value of different forms of historical practice are imbricated not only in questions about the sites and subjects of significance to historical enquiry, but also in underlying beliefs about, on the one hand, the role of emotion and imagination, and on the other, the role of *reason*, as embedded within historical research methods. For example, within practices involving specialist conservation work, emotion is compartmentalised and distanced, as Thomas Yarrow has perceptively observed: 'some ways of knowing gain traction at the expense of others' (Yarrow 2019: 15), and 'the embodiments of expert knowledge engender specific roles, while other, more "personal" judgements are factored out. These views may be expressed, but go unrecorded and are not recognized as legitimate considerations in the decision-making process' (Yarrow 2017: 105; Jones and Yarrow 2013). Non-academic and non-specialist relationships to history continue to be considered more personal, idiosyncratic, imbued with 'memory and imagination' (Weisman 2011: 22). As described, heritage is also associated with memory's 'subjective, dynamic and partial nature versus "history" as a presumably objective, fully documented and complete accounting of the past' (Sather-Wagstaff 2011: 194). Indeed, academic history 'striv[es] to maintain a sense of distance from the past' to focus on a 'broad spectrum of world events outside a person's own archive of experience' – a far cry from memory's 'untrustworthy ... affective relationship to the past' (Landsberg 2004: 19).

Rigorous academic methods are also championed as a buffer against a feared creep of 'reactionary, mythologizing amnesia' and the 'lack of scruple about historical truth' associated with the coupling of 'historical reconstruction to ideological fantasy' in heritage practice (Berger 2007: 598). The urgent need is to maintain a distinction between 'evidence-based' accounts of 'knowledge expertise' and 'play or fiction' (Harvey 2015: 537), between 'intellectual rigor' and the production of history that is 'fatuously celebratory, brainlessly bland, or just plain inaccurate' (Dresser 2010: 59) – the latter comment targeted at government-backed community outreach initiatives with which historians are increasingly compelled to engage.

These responses are bound up with claims to expertise. Hilda Kean describes how the historian, 'usually seen as professionally trained', is perceived as performing 'an active role and the "public" a passive one. The onus therefore is upon the historian to ensure that the body of knowledge transmitted is accessible. This has the dual effect of engaging "the public" but also of enhancing the separate status of the historian as the disseminator who possesses not only knowledge but the skill of transmission' (Kean 2010: 26). This particular model of public history is illustrated by the MA in Public History at Royal Holloway, University of London, which until recently boasted on its website how it 'draws very much on "in house" expertise within the department', where its 'historians continue to have a considerable impact on the public understanding of the past'. In an interview with its founder, Justin Champion, public history is deemed to draw on 'all the skills of a historian: original research among dusty papers, piecing together the story, placing it in the context of the past, interpreting it for the present. But public history also makes a special effort to bring history alive for an audience, which may well combine interest with sketchy historical knowledge and barely conscious historical preconceptions or even prejudices' (Prest 2009). Champion is quoted in the article as responding: 'Many historians don't even think about these groups'.

The notion of such an audience's 'barely conscious historical preconceptions or even prejudices' rather hints at the attitude in which some academic historians have undertaken their responsibility to engage with their 'publics'. Kean (2010: 26) calls for a different view of public history, one which 'places less emphasis on any distinctiveness of "historian" and "public" and more upon the process of how the past becomes history'. Describing the relationship between academic and popular historical practices, archaeologist John Schofield (2014: 1) concedes, in turn, that people's 'multiple views of heritage', can 'take us beyond the conventional boundaries of heritage ... beyond its comfort zone, from the special and the exceptional places and things, to the everyday'. But this, in turn, returns us to an assumption that the mundane spaces of the 'everyday' are *the* sites of public engagement with the past (becoming, for some, the sites in which to engage *with* the public *about* the past).

The historical imagination

Such an engagement with the 'public' has, however, led some historians to acknowledge how far history itself is an 'imaginative way of knowing' (Harvey 2015: 537). Work has emerged exploring, for example, emotional responses to 'doing' academic history, with Katherine Hepworth (2016: 283) pointing out how 'all people, including historians, are subject to a myriad of both conscious and unconscious influences on a daily, let alone moment by moment, basis. Whether or not we are aware of them, these influences affect our behavior, observations and thoughts'. Karen Barclay (2018: 467) describes what she deems the *positive* benefit of emotional engagement in bringing 'the past into the present', arguing: 'If acknowledging that historians feel emotions whilst doing research is hardly new, the critical capacities of such emotions are under-explored, particularly for those of us who work with the dead ... from the work on "archive fever", to subjectivity and research ethics, to affective memory, to histories of reading and mourning', emotional responses are a 'productive contribution to historical knowledge-making'.

The shift in the language used to describe the historical research process includes the notion of 'presencing' and metaphors of 'conversation' with the past, influenced by the 'rejection of the idea of objectivity' (Pihlainen 2014: 575). But this form of positioning the past 'as if indeed somehow an active agent, able to converse with the historian in meaningful ways' has also been critiqued: 'the particularity of the past as past seems to deny that possibility' and the communicative metaphor draws on a 'distinction between the creative imagination and any real access to the "otherness" of the past ... at stake in this debate is the capacity of the past to intervene on our understandings in any (disruptive) way'.

These discussions about of the role of imagination and emotion in historical research emerge out of the common use of the term 'historical imagination' in published papers. Although rarely articulated, this suggests a continuing influence of twentieth century historian R. G. Collingwood (1994 [1946]) who acknowledged – but perhaps wished to tame – the imagination, tethering it to the 'disciplined' project of working with historical sources to turn them 'into knowledge' – the imagination a 'way we construct that knowledge as about something going beyond ourselves and our social world' (Smith 2000: 105; Lemisko 2004; Smith 2007; Tosh 2010). This form of historical imagination is useful – in the Foucauldian sense – as a means of examining the 'conditions of knowledge and power in the present' (ibid: 107). But it assumes a narrow rendering of the scope of 'imaginary' as a means of reaching the 'real' – imagination here a necessary evil of historical enquiry, needing to be harnessed and controlled through rigorous methods.[12]

From a different direction, those who, as described, believe that a form of empathetic connecting with past others is an ethical imperative, or wish to take history out of the academy, tend to bolster their arguments by using the language of 'imagination' as a means of pointing to the limitations of historical 'knowledge'. Raphael Samuel, for example, observed that the 'immaculate conception of knowledge – with its refusal to countenance any traffic between the imaginary and the real – is impossible, in practice, to sustain' (Samuel 2012 [1994]: 431). Others suggest the importance of imaginative engagements to memory and place. Kitson and McHugh (2015: 504) argue that 'the intellectual logics of historic preservation are insufficient to apprehend sensual practices arising from … amorphous and ephemeral experiences' which affirms the 'ethical insufficiency of the intellect' in 'creating meaningful places'. Graham Dawson (2005: 151) has described the importance of 'imaginative meanings and associations attached to a place through storytelling or practices of remembrance [which] enable a community of people to orient themselves within and inhabit that place', and Avery Gordon described being haunted by the figure of a ghost who 'draws us affectively, sometimes against our will and always a bit magically, into the structure of feeling of a reality we come to experience, *not as cold knowledge, but as a transformative recognition*' (1997: 8; my italics). But whereas the binary between imagination and knowledge appears to be reworked in the service of arguments for particular forms of engagement with the past and empathy towards forgotten others, elsewhere there is also acknowledgement of the more complicated relationships between 'individual and social memory, between real and imaginary, event and fantasy, history and myth' (Hodgkin and Radstone 2003: 16).

These debates tend to refer to the 'public' or the 'people' as a homogeneous, undifferentiated category, and ignore how different 'publics' encounter the past in different ways, as well as the potential of people's ability to have their *own* awareness of their responses to the past (Waterton 2005). This book seeks to explore these relationships between the 'imaginary' and the 'real' within the situated experiences of the home, extending our understanding of how historical knowledge is produced and encountered in relation to, or beyond, the established technologies of historical authenticity and authority. It allows space to explore how people *themselves* might grapple with how to define the 'past' and what to value of it, of how ideas of expert distance and objective rigour might inform more emotional and imaginative responses to the pasts of their homes, and the role of 'embodied aspects' of heritage experience, and the 'affective relationships we have with our pasts' (Tolia-Kelly *et al.* 2017: 1; Atkinson 2003; Buchli and Lucas 2001).

Imagination and belief

Exploring the 'presence' of the past at home highlights the difference between traditional archaeological investigations into 'pre-habitation' of once-inhabited places where traces of habitation remain, and the 'archaeologies' of contemporary homes which continue to be *occupied*. For the latter, the past continues to be experienced, and therefore the meanings given to domestic dwelling might be shaped by an awareness of this past in particular ways. The home becomes a space for the making of histories which continually evolve, and where what gets communicated from one resident to the next, like all forms of knowledge and tradition, is 'not handed on unchanged from generation to generation but ... generated in the course of lived experience' (Nash 2008: 13). The forms of historical knowledge and practice that are prompted and informed by and result from awareness of the past at home might, albeit unconventionally, be described as forms of 'living heritage' practice.

Thus, considering the home as a site of heritage suggests the need to explore the complexity of people's responses beyond any too narrow formula for judging the value of heritage and what is worthy of being 'safeguarded' for the future. UNESCO's categorisation of 'intangible cultural heritage' signalled an important shift from a focus on artefacts to people's knowledge and skills – from an earlier model supporting 'scholars and institutions to document and preserve a record of disappearing traditions' to one which 'seeks to sustain a living, if endangered, tradition by supporting the conditions necessary for cultural reproduction' (Kirshenblatt-Gimblett 2004: 52; Foster and Gilman 2015). But early in this process, Brenda Kirshenblatt-Gimblett rightly expressed concerns about the 'arbitrariness' of the separate categories of what constituted 'intangible' and 'tangible' heritage and the processes for deciding what should be included in the inventories. She pointed, for example, to the fact that 'intangible heritage ... is not only embodied, but also inseparable from the material and social worlds of persons' (ibid: 53). What if we were to extend this 'intangible' category further to include more fluid encounters with heritage – feelings, emotions, affects, atmospheres and beliefs about the 'presence' of the past? How do you quantify, categorise and value those responses which inform how experiences of 'heritage' have impact? The official definition of 'intangible' heritage becomes irrelevant in the light of such a project. Indeed, heritage guidelines increasingly acknowledge the need to include 'subjective qualities associated with historic objects, buildings, and places, such as atmosphere, spirituality, feeling, and so forth' (Jones 2010: 141) – particularly in the context of challenges from 'alternative perspectives, in particular indigenous and non-Western approaches to heritage' (135). There have been, indeed, calls to 'integrate these intangible and subjective qualities further in conservation practice, where they tend to be marginalized' (41).

How might such 'intangible and subjective qualities' emerging from encounters with the past at home be categorised as forms of 'heritage'? Arguably, for example, any investigation of such a past (in a Western, contemporary context) needs to be open to the experiential impact of beliefs about how the past may be 'present' in different ways. We might term this a form of 'spectral heritage' which has the potential to speak back to those official heritage discourses (Smith 2004) which continue to impede or close down particular engagements with the past. The historian Karl Bell (2016 online; 2012) argues that the ghost story is one form of 'guerrilla heritage' which challenges 'systematic' forms of knowledge and 'draw(s) attention to stories erased by national memories'. Sheila Watson (2018; online) also notes that 'popular feelings about heritage such as those relating to the supernatural, are rarely taken into consideration when interpretive strategies are formulated by official bodies that tend to adopt a positivist and empirical approach to assessing the significance of historical sites, objects and monuments'. And yet, she adds, 'evidence as to how communities engage with the past is accessible in the form of locally produced histories, folk tales, popular narratives and types of local associations, and an examination of these suggest that official and private meaning making are sometimes very different from each other'. We might consider what certain stories or anecdotes (Michael 2012), circulating locally, reveal of particular histories and affective economies that would otherwise remain hidden. If the past 'becomes more important *because* it is seen to be threatened' (Yarrow 2019: 18), what place for such ad hoc experiences and forms of incomplete storytelling about unknowable things?

The agency of objects

Daniel Miller (2008: 119), explaining the belief in the agency of *homes,* points to the 'very longevity of homes and material culture' which may 'create a sense that agency lies in these things rather than in the relatively transient persons who occupy or own them'. Thomas Yarrow (2019: 12) also reflects on the belief in homes as 'pre-existing agents' – full of '"atmosphere", "character" and "personality" … quasi persons, entities that have metaphorically human capacities'. He suggests this relates to how buildings are 'sensed and experienced through practices of daily domestic life'. A home's agency is here 'linked to myriad ways of registering the past [including] acts of memory and imagination that connect people back to their own and others' lives', but also, in turn, the way homes affect – they *act on* – inhabitants, who 'acquire specific capacities through their inhabitation of these spaces'.

But despite a recent shift to consider more-than human affects and the theorising of materiality as lively, excessive and emergent (Edensor 2012; Anderson and Wylie 2009; Dewsbury *et al.* 2002; Doel 2004; Kearnes 2003), objects are never allowed to be *too* animate in these accounts. Alfred Gell's influential reflection on an object's agency relates this to social action which 'excludes … factors that make it possible … to conceive of certain objects as having agency in themselves' (Morphy 2009: 14; Fowler and Harris 2015; Gell 1998). Reacting to a traditional focus on past objects' 'origins, historical validity, material integrity' (Jones 2010: 140), archaeologist Sian Jones suggested that objects' authenticity is 'negotiated as a magical, almost numinous, quality … a form of magical communion, in a contagious sense' during the process of 'physical contact or intimate experience' of objects (143–144). The importance placed here is on touch, the haptic encounter with past things by which 'the distinction between past and present is momentarily dissolved' (Harries 2017: 110). Jones, however, suggests that such an experience and its 'powerful impact' is related to objects' 'cultural biographies' or 'social lives' (144) rather than anything innate within the objects themselves. In turn,

Cornelius Holtorf's elaboration of the notion of 'pastness' suggests that past objects are always 'cultural constructs of the present' – an outgrowth of people's perceptions, experiences and emotional engagements in a given cultural context (Holtorf 2013: 431). Responding to criticisms that this ignores the importance of objects' material presence, he concedes that objects may be granted 'pastness' due to 'evidenced knowledge of age', or to an *appearance* of age, its 'wear and tear' or 'patina' (432), concluding: 'It is no longer anything materially "inherent" which gives a historic object authenticity'.

In a recent exploration of agency (Lipman 2019), I drew upon Mitch Rose's critique of material culture theory's 'profoundly secular' forms (Rose 2011: 110), and noted a recent return to interest in Jane Bennett's (2001) suggestion that 'through certain engagements with materialities people can be *enchanted*' (Burrell 2011: 143; Woodyer and Geoghegan 2012, 2014). This suggests a different approach, one which acknowledges the 'sheer emotional, imaginative and tangible power that objects provoke around them, and asserts people's openness to the different and unusual' (ibid: 144). Bennett's concept of enchantment, some argue, 'recognises qualities of liveliness as internal rather than supplementary to objects, helping move beyond the assumption that objects simply await enlivening by human subjects' (Ramsay 2009: 198). This turn back to agency as an intrinsic quality stands in contradistinction to the continuing influence of Sigmund Freud's (2003 [1919]) belief that granting agency to *things* is a slide back into animistic, 'primitive' beliefs. But the agency of past objects is often construed in relation to their unknowability. The psychologist Ernst Jentsch described the uncanny or *unheimlich* as 'doubt as to the animate or inanimate state of things' (Jentsch 1906, in translation). Freud paraphrased this as 'intellectual uncertainty', something which is 'frightening precisely because it is not known and familiar' (Freud 2003 [1919]: 220). The assumption of, and fascination for, the unknowability of past objects continues to draw commentary. Tonya Davidson (2009: 339), for example, argues that the material traces of the past are like 'scars on the house-as-body, marking and insisting remembrance'. Caitlin DeSilvey describes how 'every object left to rot in a dark shed or an airless attic once occupied a place in an active web of social and material relations' (DeSilvey 2007: 403). As a curator, she felt a 'rift' or 'gap' between her 'desire to recover a trace of the past and the undeniable inaccessibility of that past' (417). Michael Shanks also described the process of archaeological discovery as having 'immediate contact of a sort with its original owner' but 'this is as close as we can get. It is the power of the edge between ourselves and the past, or rather an *other*; it is both proximity and distance' (Shanks 1992: 59; see also Moshenska 2006). He added that the 'raw existence of the past is impenetrable ... the particularity of what I find is fascinating, unsayable, uncanny. It is discovery, uncovering what was hidden, showing our homely and familiar categories and understanding to be insufficient' (114). The material traces of those now-absent others of the home's past likewise act to 'presence' them, for 'while absence is matter out of place, it is still placed through matter' (Meyer 2012: 109). For others it is the 'combination of the familiar and the unfamiliar' within the home which makes it uncanny (Gelder and Jacobs, 1998: 23); the spectres of the 'archaeological imagination', are 'at once horrifying and comforting' (Buchli and Lucas, 2001: 11–12). Considering how people respond to inherited objects in their home – objects which may not be 'known or familiar' – suggests we pay attention to the affective, emotional and imaginative charge they might emit. Do such objects always elicit an uncanny response, like the *memories* not their own that they embody? How do such objects act as a focus for self-other relationships within the home?

The debates about the power of past objects to affect us – and contested beliefs about their agential powers – are useful for reflecting on how people encounter past materialities in the home and how they might respond to them. But this also signals a need to take seriously broader processes of seeking 'knowledge', including the way imaginative, spectral, spiritual and emotional responses to the past at home might in themselves constitute forms of 'knowledge-making'. Although this focus does not preclude an exploration of the wider social contexts in which such responses emerge, it does caution us to avoid dismissing certain beliefs or experiences as *in*authentic, or explaining away 'false' beliefs by reducing them *only* to wider contexts (Lipman 2014; Pile 2005; Holloway 2006; Ferber 2006; Buttimer 2006). The different ways in which people may encounter and respond to the past of their homes may take us beyond narrow configurations of what we 'know' and the different ways 'knowledge' of the home's past is arrived at, contested or legitimised.

Heritage in the Home is based on qualitative research involving 35 households in England. Participants were recruited mainly through printed postcards which asked broadly: 'Have you ever wondered about the history of your home? Who lived there before you? Or the way the past has left its imprint in different ways?' These cards were deposited in borough and city archives and local history centres, to attract people who were engaged in house history research – but also in a wide range of public spaces, including noticeboards in cafes, community centres, supermarkets and libraries, in order to capture experiences of people who were not researching their home's history. Participants, whose names I have anonymised, were also recruited through workshops and public talks organised at the Geffrye Museum of the Home in east London, where I co-curated an exhibition and organised workshops. I decided to interview everyone who responded to the advert and were willing to allow me access to their homes.

As a piece of qualitative research focused on a select number of in-depth case studies, the project was not designed to be representative of all possible experiences of home and did not seek to target any particular interest or identity group. However effort was made to reach a diversity of people in order to capture a variety of experiences. To this end, I sought out community centres for particular ethnic and cultural groups, and targeted a wide range of locations with different socio-economic profiles, tenures, housing types and locations. Despite respondents ranging in terms of age, gender and class, a limitation of recruitment was that the majority were white British people, thus offering voice to particular standpoints whilst precluding others. The age of homes ranged from over 300 years old to a 1980s council apartment. Two thirds of participants were home owners and a third social or private tenants. A majority of participants had lived in their homes for some years and were mainly based in urban and suburban London (in part reflecting the success of recruitment at the Geffrye Museum) but there were case studies from other cities, market towns, villages and one in a remote rural hamlet.

I undertook basic census research for each participant, bringing copies with me to offer them at the end of my visits, in part so that I could observe their reaction to the information I had discovered about past residents. The research was conducted mainly through home visits combining semi-structured interviews with more fluid 'guided tours', which I asked participants themselves to lead. The latter offered a glimpse into immediate, embodied engagements with home, particularly its spaces and material cultures (Daniels 2001; Hitchings 2003). The former is core to the conventional qualitative approach, which, in contrast to the 'concern to flatten [the] anthropocentric hierarchy' (Lipman 2014: 20) suggests that 'it is only through our engagement with the social that we can speak of the others' (22). Kathy

Figure 1.1 A carved wooden pig found under floorboards reflecting a room's previous use by children.

Burrell (2016: 1604), in using this narrative-focused approach, describes the need for an 'almost customary' defence of the use of interviews in this '(now post) non-representational theory age', to assert that 'people *can talk* about their practices'. She correctly points to the 'moral deficit of decentring the human too much, and the obvious impossibility of academics, as humans, ever forming research *insights* which are not themselves subject based', adding that there is also 'an argument to be made for the moral weight of narratives and storytelling, for the empathy and understanding they can promote on a human level' (1605). To this end, it was important to make as much space as possible to allow people's stories to be heard, through detailed examples and the liberal use of direct quotations. With so many case studies, inevitably some voices are heard more clearly than others, and the comparative and thematic approach necessarily emphasises my own interpretation to the materials gathered – which, of course, this remains.

Heritage in the Home is divided into three parts, each with two chapters. Part I examines people's ideas of knowledge, imagination, emotion and belief in researching and otherwise engaging with the past of their homes. Part II examines people's encounters with the social histories of their homes, considering in turn the earlier residents and the more recent (and future) ones. Part III explores the material cultures of home, firstly in relation to inherited objects and then homemaking practices. The conclusion reflects on the broader contexts of engaging with the home's pasts, in particular considering the impacts of local change.

Notes

1 Different definitions of emotion and affect continue to be discussed. Eric Shouse (2005; online) clarified: 'Feelings are personal and biographical, emotions are social, and affects are prepersonal'. However, others have trouble with such a neat categorisation. Whereas affect is seen as 'highlighting the embodied state and the initial registering of events in bodies and minds', whilst emotion 'refers to the processing and packaging of affect in familiar cultural categories such as anger, grief', some caution that the two are 'flowing, dynamic, recursive and profoundly contextual, challenging static and neat formulations' (Wetherell *et al.* 2018: 1). Specifically, and most usefully for our purposes, is the argument that paying attention to emotion and affect 'allows us to deepen our understanding of how people develop attachments and commitments to the past, things, beliefs, places, traditions and institutions', as well as 'reveal the nuances of how people negotiate various forms of identity, sense of social and physical place, and feelings of wellbeing and discomfort' (2).

2 Second-hand objects can also be imaginatively adapted, as in one migrant artist's project exploring the fragmented stories of salvaged items he had bought for his new home, a flat in a council estate in south London. He described 'borrowing' their incomplete biographies as a way to 're-story the space, as well as reshape his own sense of home and identity' (Lipman and Sheringham 2016: 53; Balthazar 2016; Digby 2006). These objects were considered imbued with histories which,

in turn, were 'retrieved, usurped, and taken on as raw material for his life'. This attempt to recycle local materials to create a sense of identity in a new environment reflects ideas about the way that objects are accumulated to create or reinforce senses of belonging to place, by 'continuously re-presenting ourselves to ourselves, and telling the stories of our lives in ways which would be impossible otherwise' (Pearce 1992: 47; Strangleman 2013).

3 My German mother-in-law offers the example of a *secret recipe* as having a *heimlich* quality. This notion of the past being hidden and attendant archaeological metaphors concerning history as something buried and needing to be revealed (utilised by Freud himself), have been critiqued as reflecting an outmoded 'salvage paradigm' which, it is argued, no longer reflects heritage practice which should instead embrace surfaces and futures (Harrison 2011: 13; Harrison 2013; MacDonald 2011; see Chapter 5).

4 Niche markets have developed to play on such anxieties. In North America you can pay for research into a home's history prior to purchase to make sure it is unlikely to be 'haunted'. Estate agents in Hong Kong reflect belief in haunted homes by also offering information about the manner of past residents' deaths.

5 The biographical model is increasingly used across popular, fictional and academic writing and museum culture. The history of a locality and wider historical change is explored and conveyed through the lives of the sequence of residents in one dwelling, home becoming 'no longer just dwellings but … untold stories of lives being lived' (Berger 1984: 64. See also Blunt 2008; Burton 2003; Chee 2005; Myerson 2004).

6 A focus on emotional geographies and histories (Davidson *et al.* 2005) has included interest in geographies of 'enthusiasm' (Geoghegan 2014; Craggs *et al.* 2016; DeLyser 2016, 2011). Others draw attention to how the 'politics of difference' is 'felt, embodied, intense', co-constituting 'practices of meaning-making' (Tolia-Kelly *et al.* 2017: 3). The latter includes an argument by feminist, queer and critical race scholars that 'negative feelings' such as anger and shame are potentially productive – at the least, that they 'should not be engaged with in terms of pathology' (Dragojlovik 2018: 99). Elsewhere, many heritage theorists have focused on work grounded in 'difficult knowledge' (Lehrer *et al.* 2011; Sather-Wagstaff 2011; Schofield, Johnson and Beck 2002) – described variously as 'heritage that hurts' (Schofield *et al.*, 2002) and 'negative heritage' (Meskell 2002). However, a growing interest in the role of *enchantment* in archaeological and heritage practice could be seen in part as a reaction to a focus on negative and traumatic events. Sara Perry (2019), for example, utilises Bennett's idea of enchantment to draw out the positive potential of people's 'affective response' to archaeology (which 'can move us'), arguing for ways this might be harnessed to 'act back on the world in constructive, ethically-minded ways' (354). Archaeologists should deliberately 'foster affect' which might 'encourage genuine social action' (355) (see also Fredengren 2016 for a related argument). The relationship between enchantment, affect and empathy here might, however, downplay unintended – or more problematic – responses to being affected by the past. Denise Lawrence-Zuniga (2014: 824), for example, describes how 'preserved landscapes' are 'enchanted' in the specific sense of becoming 'naturalized, normalized, and taken for granted' and 'hegemonic for the groups who succeed in promoting them' – at the expense of others.

7 Recent work in cultural theory has tended to seek to frame experience *beyond* a subject-object (or self-other) binary, focusing instead on the 'assemblage of

things' (Deleuze and Guattari 1987), those 'adhoc groupings of diverse elements, of vibrant materials of all sorts' (Bennett 2010: 23), the 'multiple bits-and-pieces [which] accrete and align over time' (McFarlane 2011: 653). An attendant influential shift has described a particular rendering of affect as a form of emergent intensity, the world in a state of *becoming* rather than *being* – part of the project to decentre notions of fixed identity when considering the 'object' of self, other and place and seek creative methods to explore the relational complexity of things. But this focus has been critiqued for its 'presentist' fixation, ignoring the 'importance of history in shaping the future' (Uprichard 2012: 134) and the role of personal and collective memory in shaping the present (Callard and Papoulias 2010: 247). Increasingly, affect theorists acknowledge how 'specific historical contingencies' inform how affects are 'experienced and qualified' (Dragojlovic 2015: 332; Tolia-Kelly 2016; Crang and Tolia-Kelly 2010; Thien 2005). Sensations experienced at moments of encounter between bodies, objects and places are seen to be *already* mediated; our reading of feelings are 'tied to a past history of readings ... the process of recognition ... is bound up with what we already know' (Ahmed 2004b: 25). Affects emerge from both immediate encounters and the 'geo-historicity of the body' (Anderson 2014: 167) as each encounter contains reference to 'past encounters [and is] made through accumulated relations, dispositions and habits'. This idea has also drawn attention to the way affective encounters are mediated by broader processes, 'inseparable from other times and space', including the way 'past encounters between coloniser and colonised' dictate the 'affects of race'.

8 The gender divide was discernible in my previous research exploring experiences of the domestic uncanny. In a number of cases, husbands assumed that their wives were more suggestible and emotional and therefore more susceptible to experiencing and believing in the supernatural (Lipman 2014; Blackman and Walkerdine 2001).

9 The focus on the situated experiences of affect through the prism of 'atmosphere' is useful for considering practices and experiences of affective encounters at *particular spaces*, including homes, extending a current focus on 'conventional' sites of heritage. These are often designed or managed 'to make visitors "feel the era", to facilitate a physical and emotional sense of the ... experience that is historically authentic' (Crouch 2015). Theories of 'atmosphere' tend to assume they are 'attached' to places (an 'unpleasant or unhappy atmosphere may persist' [Ackroyd 2000: 271]), to focus on sensations which get into us on a physiological level (Brennan 2004), as well as those we ourselves *bring* to places. Ahmed's emphasis on how emotions circulate and affect's capacity to be 'contagious' (Thrift 2004) are considered more helpful in avoiding the idea that atmospheres are 'simply "received" by a neutral body' (Anderson 2014, 2009). This also focuses attention on 'situatedness and immersion within an environment rather than a hyperactive world of endless flows and relations'. Thus, atmosphere as a concept has been taken up as a 'way of emplacing affect and affect theory' (Brown *et al.* 2019), including exploring how it 'manifest[s] as we dwell in or move through our surroundings' (Sumartojo and Pink 2019). Attending to the conditions in which atmospheres emerge and the 'meanings that people ascribe to them' allows us to explore how people 'shape their understandings of their experiences'.

10 Beyond a focus on the 'intersubjective, performative, and contingent aspects of the formation of a subject' (Korf 2008: 715), some have argued that the self also possesses 'a degree of centring and consistency within the relations that shape it' (Conradson 2005: 104–105), and that an individual's 'capacity to act requires

some degree of coherence in the conception of self' (Varley 2008: 55; Blackman *et al.* 2008).

11 Landsberg's theory has been described in the light of a 'turn' to ethics in theorising encounters and engagements with others and attunement to the 'difference that characterizes the self-other relation' (Abel 2006: 377). As Anderson (2014) also argues, the recent interest in enchantment (Bennett 2001) and generosity (Diprose 2002) also reflect a concern to theorise the ethics of attachment to people and things.

12 The question of the place of the imagination in historical enquiry – and how 'imagination' is to be defined – is not new, but signals a broader debate about the nature of history as an academic discipline. This has focused recently on its social responsibilities, and on the 'construction of meaning in historical discourse and hence also on the ideological and political consequences of doing history ... shoring up the integrity of *historical* knowledge and separating it from more general talk about the significance of the past' (Pihlainen 2016: 144). One focus for this discussion has been Michael Oakeshott's idea of 'the practical past' (Oakeshott 2004), as distinguished from 'what is often viewed as its detached other, "the historical past"' (Pihlainen 2016: 145. See also Munslow 2014, 1997; LaCapra 2004; Jenkins 2003 [1991]; Fulbrook 2002; Paul 2011). Pihlainen wishes to argue that there is 'some inescapable connection that historians (and perhaps people in general) have with the historical past, the past beyond their subjective experience, which is present in their daily lives'.

References

Abel, M (2006): 'A Review of "Prosthetic Memory: The Transformation of American Remembrance in the Age of Mass Culture"'. *Quarterly Review of Film and Video* 23.4: 377–388.

Abraham, N and Torok, M (1994): *The Shell and the Kernel: Renewals of Psychoanalysis, volume 1*. Chicago: University of Chicago Press.

Ackroyd, P (2000): *London: The Biography*. London: Chatto & Windus.

Ahmed, S (2000): *Strange Encounters: Embodied Others in Post-Coloniality*. London: Routledge.

Ahmed, S (2004a): 'Affective Economies'. *Social Text* 79, 22.2: 117–139.

Ahmed, S (2004b): *The Cultural Politics of Emotion*. Edinburgh: Edinburgh University Press.

Ahmed, S and Stacey, J (2001): 'Introduction: Dermographies'. In S Ahmed and J Stacey (eds.) *Thinking Through the Skin*. London: Routledge: 1–18.

Al-Ali, N and Koser, K (2002): *Transnational Communities and the Transformation of Home*. London: Routledge.

Anderson, B (2009): 'Affective Atmospheres'. *Emotion, Space and Society* 2.2: 77–81.

Anderson, B (2014): *Encountering Affect: Capacities, Apparatuses, Conditions*. London: Ashgate.

Anderson, B and Wylie, J (2009): 'On Geography and Materiality'. *Environment and Planning A* (41): 318–335.

Ashton, P (2010): 'Introduction: Going Public'. *Public History Review* 17: 1–15.

Ashton, P and Kean, H (eds.) (2009): *People and Their Pasts: Public History Today*. Basingstoke: Palgrave Macmillan.

Atkinson, A (2003): 'Heritage, Self and Place'. *ACH: Australian Cultural History* 23: 161–172.

Atkinson, D (2008): 'The Heritage of Mundane Places'. In B Graham and P Howard (eds.) *The Ashgate Research Companion to Heritage and Identity*. Farnham: Ashgate: 381–396.

Austin D, Dowdy M, and Miller, J (1997): *Be Your Own House Detective*. London: BBC Books.

Bachelard, G (1994 [1958]): *The Poetics of Space*. Boston: Beacon Press.

Backe-Hansen, M (2011): *House Histories: The Secrets Behind your Front Door*. Stroud: The History Press.

Balthazar, A (2016): 'Old Things with Character: The Fetishization of Objects in Margate, UK'. *Journal of Material Culture* 21.4: 448–464.

Bammer, A (1992): 'Editorial: The Question of "Home"'. *New Formations: A Journal of Culture/Theory/Politics* 17: vii–xi.

Barclay, K (2018): 'Falling in Love with the Dead'. *Rethinking History* 22.4: 450–473.

Barratt, (2001): *Tracing the History of your Home*. Richmond, Surrey: Public Record Office.

Bell, K (2012): *The Magical Imagination: Magic and Modernity in Urban England 1780–1914*. Cambridge: Cambridge University Press.

Bell, K (2016): 'Inheriting Phantasmal Cities: Urban Ghost Lore as Intangible Heritage'. Abstract from conference paper, available at: https://inheritingthecity. files.wordpress.com/2015/11/25_dr-karl-bell_inheriting-phantsmal-cities-urban-ghost-lore-as-intangible-heritage.pdf.

Benjamin, W (2002 [1935]): *The Arcades Project*. Cambridge MA: Harvard University Press.

Bennett, J (2001): *The Enchantment of Modern Life: Attachments, Crossings, and Ethics*. Princeton, NJ: Princeton University Press.

Bennett, J (2010): *Vibrant Matter: A Political Ecology of Things*. Durham, NC: Duke University Press.

Berger, J (1984): *And Our Faces, My Heart, Brief as Photos*. London: Bloomsbury.

Berger, J (2007): 'Which Prosthetic? Mass Media, Narrative, Empathy, and Progressive Politics'. *Rethinking History* 11.4: 597–612.

Blackman, L, Cromby, J, Hook, D, Papadopoulos, D and Walkerdine, V (2008): 'Editorial: Creating Subjectivities'. *Subjectivities* 22.1: 1–27.

Blackman, L and Walkerdine, V (2001): *Mass Hysteria: Critical Psychology and Media Studies*. Basingstoke: Palgrave.

Blakely, H and Moles, K (2019): 'Everyday Practices of Memory: Authenticity, Value and the Gift'. *Sociological Review* 67.3: 621–634.

Blunt, A (2008): 'The Skyscraper Settlement: Home and Residence at Christodora House'. *Environment and Planning A* 40: 550–571.

Blunt, A and Bonnerjee, J (2013): 'Home, City and Diaspora: Anglo–Indian and Chinese Attachments to Calcutta'. *Global Networks* 13.2: 220–240.

Blunt, A and Dowling, R (2006): *Home*. Abingdon: Routledge.

Blunt, A and Sheringham, O (2018): 'Home-City Geographies: Urban Dwelling and Mobility'. *Progress in Human Geography* 43.5: 815–834.

Bourdieu, P (1977): *Outline of a Theory of Practice*. Cambridge: Cambridge University Press.

Brennan, T (2004): *The Transmission of Affect*. Ithaca, NY: Cornell University Press.

Brogan, K (1998): *Cultural Haunting: Ghosts and Ethnicity in Recent American Literature*. Charlottesville: University of Virginia.

Brooks, P (2007): *How to Research Your House*. Oxford: How To Books

Brown, S, Kanyeredzi, A, McGrath, L, Reavey, P and Tucker, I (2019): 'Affect Theory and the Concept of Atmosphere'. *Distinktion: Journal of Social Theory* 20.1: 5–24.

Buchli, V and Lucas, G (eds.) (2001): *Archaeologies of the Contemporary Past*. London: Routledge.

Burrell, K (2016): 'Lost in the "Churn"? Locating Neighbourliness in a Transient Neighbourhood'. *Environment & Planning A* 48.8: 1599–1616.

Burrell, K (2011): 'The Enchantment of Western Things: Children's Material Encounters in Late Socialist Poland'. *Transactions of the Institute of British Geographers* 36: 143–156.

Burton, A (2003): *Dwelling in the Archive: Women Writing House, Home and History in Late Colonial India*. Oxford: Oxford University Press.

Bushell, P (1989): *Tracing the History of Your House*. London: Pavilion Books.

Buttimer, A (2006): 'Afterword: Reflections on Geography, Religion, and Belief Systems'. *Annals of the Association of American Geographers* 96.1: 197–202.

Byrne, D (2013): 'Love & Loss in the 1960s'. *International Journal of Heritage Studies* 19.6: 596–609.

Callard, F and Papoulias, C (2010): 'Affect and Embodiment'. In S Radstone and B Schwarz (eds.) *Memories: Histories, Theories, Debates*. New York: Fordham University Press.

Cannell, F (2011): 'English Ancestors: The Moral Possibilities of Popular Genealogy'. *Journal of the Royal Anthropological Institute* 17: 462–480.

Chapman, T and Hockey, J (eds.) (1999): *Ideal Homes? Social Change and Domestic Life*. London: Routledge.

Chee, L (2005): 'An Architecture of Twenty Words: Intimate Details of a London Blue Plaque House'. In H Heynen and G Baydar (eds.) *Negotiating Domesticity: Spatial Productions of Gender in Modern Architecture*. London: Routledge: 181–195.

Cloke, P and Jones, O (2004): 'Turning in the Graveyard: Trees and the Hybrid Geographies of Dwelling, Monitoring and Resistance in a Bristol Cemetery'. *Cultural Geographies* 11: 313–341.

Collingwood, R G (1994 [1946]): *The Idea of History: Revised Edition with Lectures 1926–1928*. Oxford: Clarendon Press.

Connerton, P (1989): *How Societies Remember*. Cambridge: Cambridge University Press.

Conradson, D (2005): 'Freedom, Space and Perspective: Moving Encounters with other Ecologies'. In J Davidson, L Bondi and M Smith (eds.) *Emotional Geographies*. Aldershot: Ashgate: 103–116.

Craggs, R, Geoghegan, H and Neate, H (2016): 'Managing Enthusiasm: Between 'Extremist' Volunteers and 'Rational' Professional Practices in Architectural Conservation'. *Geoforum* 74: 1–8.

Crang, M and Tolia-Kelly, D (2010): 'Nation, Race and Affect: Senses and Sensibilities at National Heritage Sites'. *Environment and Planning A* 42: 2315–2331.

Cresswell, T (2004): *Place: A Short Introduction*. Oxford: Blackwell.

Cresswell, T (2012): 'Value, Gleaning and the Archive at Maxwell Street, Chicago'. *Transactions of the Institute of British Geographers* 37.1: 164–176.

Crouch, D (2015): 'Affect, Heritage, Feeling'. In E Waterton and S Watson, (eds.) *The Palgrave Handbook of Contemporary Heritage Research*. London: Palgrave Macmillan: 177–190.

Daniels, I (2001): 'The "Untidy" Japanese Home'. In Miller, D (ed.) *Home Possessions: Material Culture Behind Closed Doors*. Oxford: Berg: 201–230.

Davidson, J, Bondi, L, and Smith, M (eds.) (2005): *Emotional Geographies*. Abingdon: Routledge.

Davidson, T (2009): 'The Role of Domestic Architecture in the Structuring of Memory'. *Space and Culture* 12.3: 332–342.

Dawson, G (2005): 'Trauma, Place and the Politics of Memory: Bloody Sunday, Derry, 1972–2004'. *History Workshop Journal* 59: 151–178.

De Certeau, M (1984 [2011]): *The Practice of Everyday Life*. Berkeley: University of California Press.

De Certeau, M (1998): *The Practices of Everyday Life, volume 2: Living and Cooking*. Minneapolis: University of Minnesota Press.

Deleuze, G and Guattari, F (trans. Massumi, B) (1987): *A Thousand Plateaus: Capitalism and Schizophrenia*. Minneapolis: University of Minnesota Press.

DeLyser, D (2011): 'Participatory Historical Geography? Shaping and Failing to Shape Social Memory at an Oklahoma Memorial'. In S Daniels, D DeLyser, D Entrikin, and D Richardson (eds.): *Envisioning Landscapes, Making Worlds: Geography and the Humanities*. London: Routledge: 177–187.

DeLyser, D (2016): 'Careful Work: Building Public Cultural Geographies'. *Social & Cultural Geography* 17.6: 808–812.

DeSilvey, C (2006): 'Observed Decay: Telling Stories with Mutable Things'. *Journal of Material Culture* 11.3: 318–338.

DeSilvey, C (2007): 'Salvage Memory: Constellating Material Histories on a Hardscrabble Homestead'. *Cultural Geographies* 14: 401–424.

Dewsbury, J, Harrison, P, Rose, M, and Wylie, J (2002): 'Enacting Geographies'. *Geoforum* 32.4: 437–440.

Dicks, B (2000): 'Encoding and Decoding the People: Circuits of Communication at a Local Heritage Museum'. *European Journal of Communication* 15.1: 61–78.

Digby, S (2006): 'The Casket of Magic: Home and Identity from Salvaged Objects'. *Home Cultures* 3.2: 169–190.

Diprose, R (2002): *Corporeal Generosity: On Giving with Nietzsche, Merleau-Ponty, and Levinas*. New York: State University of New York Press.

Doel, M (2004): 'Poststructuralist Geographies: The Essential Selection'. In P Cloke, P Crang and M Goodwin (eds.) *Envisioning Human Geographies*. London: Edward Arnold: 146–171.

Douglas, M (1991): 'The Idea of a Home: A Kind of Space'. *Social Research* 58.1: 287–307.

Dragojlovic, A (2015): 'Affective Geographies: Intergenerational Hauntings, Bodily Affectivity and Multiracial Subjectivities'. *Subjectivity* 8.4: 315–334.

Dragojlovic, A (2018): 'Politics of Negative Affect: Intergenerational Hauntings, Counter-Archival Practices and the Queer Memory Project'. *Subjectivity* 11: 91–107.

Dresser, M (2010): 'Politics, Populism, and Professionalism: Reflections on the Role of the Academic Historian in the Production of Public History'. *Public Historian* 32.3: 39–63.

Edensor, T (2001): 'Hauntings in the Ruins: Matter and Immateriality'. *Space and Culture* 11/12: 42–51.

Edensor, T (2012): 'Illuminated Atmospheres: Anticipating and Reproducing the Flow of Affective Experience in Blackpool'. *Environment and Planning D: Society and Space* 30: 1103–1122.

Ferber, M (2006): 'Critical Realism and Religion: Objectivity and the Insider/Outsider Problem'. *Annals of the Association of American Geographers* 96.1: 176–181.

Findlay, A (2009): 'Remaking Homes: Gender and the Representation of Place'. *Home Cultures* 6.2: 115–122.

Foster, M and Gilman, L (eds.) (2015): *UNESCO on the Ground: Local Perspectives on Intangible Cultural* Heritage. Bloomington: Indiana University Press.

Fowler, C and Harris, O (2015): 'Enduring Relations: Exploring a Paradox of New Materialism'. *Journal of Material Culture* 20.2: 127–148.

Fredengren, C (2016): 'Unexpected Encounters with Deep Time Enchantment. Bog Bodies, Crannogs and 'Otherworldly' Sites: The Materializing Powers of Disjunctures in Time'. *World Archaeology* 48.4: 482–499.

Freud, S ([1919] 2003): *The Uncanny*. London: Penguin Books.

Fulbrook, M (2002): *Historical Theory: Ways of Imagining the Past*. New York: Routledge.

Gelder, K and Jacobs, J (1998): *Uncanny Australia: Sacredness and Identity in a Postcolonial Nation*. Melbourne: Melbourne University Press.

Gell, A (1998): *Art and Agency*. Oxford: The Clarendon Press.

Geoghegan, H (2014): 'A New Pattern for Historical Geography: Working with Enthusiast Communities and Public History'. *Journal of Historical Geography* 46:105–107.

Gordon, A (1997): *Ghostly Matters: Haunting and the Sociological Imagination*. Minneapolis: University of Minnesota Press.

Gregson, N, Metcalfe, A and Crewe, L (2007): 'Moving Things Along: The Conduits and Practices of Divestment in Consumption'. *Transactions of the Institute of British Geography* 32.2: 187–200.

Grossman, A (2015): 'Forgotten Domestic Objects'. *Home Cultures: The Journal of Architecture, Design and Domestic Space* 12.3: 291–310.

Halbwachs, M (1994 [1925]): *On Collective Memory*. Chicago: University of Chicago Press.

Hamilton, P and Ashton, P (2003): 'At Home in the Past: Initial Findings from the Survey'. *Australian Cultural History* 22: 5–30.

Hanlon, J, Hostetter, E and Post, C (2011): 'Special Issue Introduction: Everyday Landscapes: Past and Present, Presence and Absence'. *Material Culture* 43.2: 1–5.

Haraway, D J (1997): *Modest_Witness@Second_Millenium.FemaleMan_Meets_ OncoMouse: Feminism and Technoscience*. New York: Routledge.

Harries, J (2017): 'A Stone that Feels Right in the Hand: Tactile Memory, the Abduction of Agency and Presence of the Past'. *Journal of Material Culture* 22.1: 110–130.

Harrison, P (2007): 'The Space between Us: Opening Remarks on the Concept of Dwelling'. *Environment and Planning D: Society and Space* 25: 625–647.

Harrison, R (2011): 'Surface Assemblages: Towards an Archaeology *in* and *of* the Present'. *Archaeological Dialogues* 18.2: 141–161.

Harrison, R (2013): *Heritage: Critical Approaches*. Abingdon: Routledge.

Harvey, D (1996): *Justice, Nature and the Geography of Difference*. Cambridge, MA: Blackwell.

Harvey, K (2015): 'Envisioning the Past: Art, Historiography and Public History'. *Cultural and Social History* 12.4: 527–54.

Hepworth, K (2016): 'History, Power and Visual Communication Artefacts'. *Rethinking History* 20.2: 280–302.

Hewison, R (1987): *The Heritage Industry*. London: Methuen.

Hill, J (2007): 'The Story of the Amulet: Locating the Enchantment of Collections'. *Journal of Material Culture* 12.1: 65–87.

Hirsch, M (2012): *The Generation of Postmemory: Writing and Visual Culture after the Holocaust.* New York: Columbia University Press.

Hitchings, R (2003): 'People, Plants and Performance: On Actor Network Theory and the Material Pleasures of the Private Garden'. *Social and Cultural Geography* 4.1: 99–113.

Hitchings, R (2004): 'At Home with Someone NonHuman'. *Home Cultures* 1.2: 169–186.

Hodgkin, K and Radstone, S (2003): *Contested Pasts: The Politics of Memory.* London: Routledge.

Holloway, J (2006): 'Enchanted Spaces: The Séance, Affect, and Geographies of Religion'. *Annals of the Association of American Geographers.* 96.1: 182–187.

Holtorf, C (2013): 'On Pastness: A Reconsideration of Materiality in Archaeological Object Authenticity'. *Anthropological Quarterly* 86.2: 427–443.

hooks, b (1990): *Yearning: Race, Gender, and Cultural Politics.* Boston: South End Press.

Hoskins, A (2003): 'Signs of the Holocaust: Exhibiting Memory in a Mediated Age'. *Media, Culture and Society* 25.1: 7–22.

Hurdley, R (2013): *Home, Materiality, Memory and Belonging: Keeping Culture.* Basingstoke: Palgrave Macmillan.

Jacobs, J and Smith, S (2008): 'Living Room: Rematerializing Home'. *Environment and Planning* A 40: 515–519.

Jeffries, N, Owens, A, Hicks, D, Featherby, R and Wehner, K (2009): 'Rematerialising Metropolitan Histories? People, Places and Things in Modern London'. In Horning, A and Palmer, M (eds.) *Crossing Paths or Sharing Tracks? Future Directions in the Archaeological Study of Post-1550 Britain and Ireland, volume 5.* Woodbridge: Boydell and Brewer.

Jenkins, K (2003 [1991]): *Re-Thinking History.* Abingdon: Routledge.

Jentsch, E (1906): 'On the Psychology of the Uncanny'. *Psychiatrisch-Neurologische Wochenschrift* 8.22.

Jesser, N (1999): 'Home and Community in Toni Morrison's Beloved'. *African American Review* 33.2: 325–345.

Jones, S (2010): 'Negotiating Authentic Objects and Authentic Selves: Beyond the Deconstruction of Authenticity'. *Journal of Material Culture* 15. 2: 181–203.

Jones, S and Yarrow, T (2013): 'Crafting Authenticity: An Ethnography of Conservation Practice'. *Journal of Material Culture* 18.1: 3–26.

Kaika, M (2004): 'Interrogating the Geographies of the Familiar: Domesticating Nature and Constructing the Autonomy of the Modern Home'. *International Journal of Urban and Regional Research* 28: 265–286.

Kean, H (2010): 'People, Historians, and Public History: Demystifying the Process of History Making'. *Public Historian* 32.3: 25–38.

Kearnes, M (2003): 'Geographies that Matter – The Rhetorical Deployment of Physicality?'. *Social & Cultural Geography* 4.2: 139–152.

Kirshenblatt-Gimblett, B (2004): 'Intangible Heritage as Metacultural Production'. *Museum International* 56.1/2: 52–65.

Kitson, J and McHugh, K (2015): 'Historic Enchantments – Materializing Nostalgia'. *Cultural Geographies* 22.3: 487–508.

Korf, B (2008): 'A Neural Turn? On the Ontology of the Geographical Subject'. *Environment and Planning A* 40: 715–732.

LaCapra, D (2004): *History in Transit: Experience, Identity, Critical Theory*. Ithaca, NY: Cornell University Press.

Ladino, J (2015): 'Mountains, Monuments, and other Matter: Environmental Affects at Manzanar'. *Environmental Humanities* 6: 131–157.

Landsberg, A (2004): *Prosthetic Memory: The Transformation of American Remembrance in the Age of Mass Culture*. New York: Columbia University Press.

Lang, R (1985): 'The Dwelling Door: Towards a Phenomenology of Transition'. In D Seamon and R Mugerauer (eds.): *Dwelling, Place and Environment*. Dordrecht: Springer.

Lawrence-Zuniga, D (2014): 'Bungalows and Mansions: White Suburbs, Immigrant Aspirations, and Aesthetic Governmentality'. *Anthropological Quarterly* 87.3: 819–854.

Lawrence-Zuniga, D (2016): *Protecting Suburban America: Gentrification, Advocacy, and the Historic Imaginary*. London: Bloomsbury.

Lehrer, E, Milton, C, and Patterson, M (eds.) (2011): *Curating Difficult Knowledge: Violent Pasts in Public Places*. New York: Palgrave Macmillan

Lemisko, L (2004): 'The Historical Imagination: Collingwood in the Classroom'. *Canadian Social Studies* 28.2.

Liddington, J (2002): 'What Is Public History? Publics and Their Pasts, Meanings and Practices'. *Oral History* 30.1: 83–93.

Lipman, C (2016 [2014]): *Co-Habiting with Ghosts: Knowledge, Experience, Belief and the Domestic Uncanny*. London: Routledge.

Lipman, C (2019): 'Living with the Past at Home: The Afterlife of Inherited Domestic Objects'. *Journal of Material Culture* 24.1: 83–100.

Lipman, C and Nash, C (2019): 'Domestic Genealogies: How People Relate to Those Who Once Lived in Their Homes'. *Cultural Geographies* 26.3: 273–288.

Lipman, C and Sheringham, O (2016): 'Restor(y)ing Home: Reflections on Stories, Objects and Space in Balin House Projects'. In T Khonsari (ed.) *My Home is Your Home*. London: Public Works Publishing: 52–61.

Lorimer, H (2005): 'Cultural Geography: The Busyness of Being "More-than-Representational"'. *Progress in Human Geography* 29.1: 83–94.

Lowenthal, D (2015 [1985]): *The Past Is Another Country: Revisited*. Cambridge: Cambridge University Press.

Lyon, D (2013): 'The Labour of Refurbishment: The Building and the Body in Space and Time'. In S Pink, D Tutt and A Dainty (eds.): *Ethnographic Research in the Construction Industry*. New York: Routledge: 23–39.

MacDonald, F (2011): 'Doomsday Fieldwork, or, How to Rescue Gaelic Culture? The Salvage Paradigm in Geography, Archaeology, and Folklore, 1955–62'. *Environment and Planning D: Society and Space* 29: 309–335.

Massey, D (1994): *Space, Place, and Gender*. Minneapolis: University of Minnesota Press.

May, J and Thrift, N (2001): 'Introduction'. In J May and N Thrift (eds.) *Timespace: Geographies of Temporality*. London: Routledge: 1–46.

May, V (2017): 'Belonging from Afar: Nostalgia, Time and Memory'. *Sociological Review* 65.2: 401–415.

McFarlane, C (2011): 'The City as Assemblage: Dwelling and Urban Space'. *Environment and Planning D: Society and Space* 29: 649–671.

Meskell, L (2002): 'Negative Heritage and Past Mastering in Archaeology'. *Anthropological Quarterly* 75.3: 557–574.

Meyer, M (2012): 'Placing and Tracing Absence: A Material Culture of the Immaterial'. *Journal of Material Culture* 17.1: 103–110.

Michael, M (2012): 'Anecdote'. In C Lury and N Wakeford (eds.) *Inventive Methods: The Happening of the Social*. London: Routledge: 25–35.

Miller, D (2001): 'Behind Closed Doors'. In D Miller (ed.) *Home Possessions: Material Culture Behind Closed Doors*. Oxford: Berg: 1–19.

Miller, D (2008): *The Comfort of Things*. Cambridge: Polity Press.

Morley, D (2000): *Home Territories: Media, Mobility and Identity*. London: Routledge.

Morphy, H (2009): 'Art as a Mode of Action: Some Problems with Gell's Art and Agency'. *Journal of Material Culture* 14.1: 5–27.

Moshenska, G (2006): 'The Archaeological Uncanny'. *Public Archaeology* 5.2: 91–99.

Moussouri, T and Vomvyla, E (2015): 'Conversations about Home, Community and Identity'. *Archaeology International* 18: 97–112.

Munslow, A (2014): 'Presence: Philosophy, History, and Cultural Theory for the Twenty-First Century'. *Rethinking History* 18.4: 633–637.

Myerson, J (2004): *Home: The Story of Everyone Who Ever Lived in Our House*. London: Flamingo.

Nash, C (2002): 'Genealogical Identities'. *Environment and Planning D: Society and Space* 20.1: 27–52.

Nash, C (2005): 'Local Histories in Northern Ireland'. *History Workshop Journal* 60: 45–68.

Nash, C (2008): *Of Irish Descent: Origin Stories, Genealogy, and the Politics of Belonging*. Syracuse: Syracuse University Press.

Nasta, S (2002): *Home Truths: Fictions of the South Asian Diaspora in Britain*. Basingstoke: Palgrave Macmillan.

Nora, P (1996): *Realms of Memory: The Construction of the French Past, Volume 1 – Conflicts and Divisions: Conflicts and Divisions v.1*. New York: Columbia University Press.

Oakeshott, M and O'Sullivan, L (2004): *What Is History? And Other Essays: Selected Writings: v.1*, edited by L. O'Sullivan. Charlottesville: Imprint Academic.

Owens, A, Jeffries, N, Wehner, K and Featherby, R (2010): 'Fragments of the Modern City: Material Culture and the Rhythms of Everyday Life in Victorian London'. *Journal of Victorian Culture* 15.2: 212–225.

Paul, H (2011): *Hayden White: The Historical Imagination*. Cambridge: Polity Press.

Pearce, S (1992): *Museums, Objects and Collections: A Cultural Study*. Leicester: Leicester University Press.

Peeren, E (2010): 'Everyday Ghosts and Ghostly Everyday in Amos Tutuola, Ben Okri, and Achille Mbembe'. In M Blanco and E Peeren (eds.) *Popular Ghosts: The Haunted Spaces of Everyday Culture*. New York: Continuum: 106–117.

Perry, S (2019): 'The Enchantment of the Archaeological Record'. *European Journal of Archaeology* 22.3: 354–371.

Pihlainen, K (2014): 'There's Just No Talking with the Past'. *Rethinking History* 18.4: 575–582.

Pihlainen, K (2016): 'Historians and "the Current Situation"'. *Rethinking History* 20.2: 143–153.

Pile, S (2005): *Real Cities: Modernity, Space and the Phantasmagorias of City Life*. London: Sage.

Pratt, G and Hanson, S (1994): 'Geography and the Construction of Difference'. *Gender, Place and Culture* 1: 5–29

Prest, M (2009): 'Can Royal Holloway's New Master's Degree Raise History from the Dead?' *The Independent Newspaper*, 15 October 2009. www.independent.co.uk/student/postgraduate/postgraduate-study/can-royal-holloways-new-masters-degree-raise-history-from-the-dead-1802639.html.

Radstone, S and Hodgkin, K (2003): 'Regimes of Memory: An Introduction'. In Radstone, S and Hodgkin, K (eds.) *Memory Cultures: Memory, Subjectivity and Recognition*. New Brunswick, NJ: Transaction Publishers.

Ramsay, N (2009): 'Taking-Place: Refracted Enchantment and the Habitual Spaces of the Tourist Souvenir'. *Social & Cultural Geography* 10.2: 197–217.

Robertson, I (2015): 'Hardscrabble Heritage: The Ruined Blackhouse and Crofting Landscape as Heritage from below'. *Landscape Research* 40.8: 993–1009.

Robertson, I (ed.) (2016 [2012]): *Heritage from Below*. New York: Routledge.

Rose, M (2011): 'Secular Materialism: A Critique of Earthly Theory'. *Journal of Material Culture* 16.2: 107–129.

Rosenzweig, R and Thelen, D (1998): *The Presence of the Past: Popular Uses of History in American Life*. New York: Columbia University Press.

Rubenstein, R (2001): *Home Matters: Longing and Belonging, Nostalgia and Mourning in Women's Fiction*. New York: Palgrave.

Samuel, R (2012 [1994]): *Theatres of Memory: Past and Present in Contemporary Culture*. London: Verso.

Sather-Wagstaff, J (2011): *Heritage That Hurts: Tourists in the Memoryscapes of September 11*. Walnut Creek, CA: West Coast Press.

Sather-Wagstaff, J (2015): 'Heritage and Memory'. In E Waterton and S Watson (eds.) *The Palgrave Handbook of Contemporary Heritage Research*. London: Palgrave Macmillan: 191–204.

Schmitz, S and Ahmed, S (2014): 'Affect/Emotion: Orientation Matters. A conversation between Sigrid Schmitz and Sara Ahmed'. *Freiburger Zeitschrift für Geschlechterstudien* 20.2: 97–108.

Schofield, J (ed.) (2014): *Who Needs Experts? Counter-Mapping Cultural Heritage*. Abingdon: Routledge.

Schofield, J, Johnson, W, and Beck, C (2002): *Matériel Culture: The Archaeology of 20th Century Conflict* London: Routledge.

Scott, S (2009): *Making Sense of Everyday Life*. Cambridge: Polity Press.

Seamon, D (1980): 'Body-Subject, Time-Space Routines and Place-Ballets'. In Buttimer, A and Seaman, D (eds.) *The Human Experience of Space and Place*. London: Croom Helm: 148–165.

Shanks, M (1992): *Experiencing the Past: On the Character of Archaeology*. London: Routledge.

Sheringham, O (2010): 'A Transnational Space? Transnational Practices, Place-Based Identity and the Making of 'Home' among Brazilians in Gort, Ireland'. *Portuguese Studies* 26.1: 60–78.

Shouse, E (2005): 'Feeling, Emotion, Affect'. *M/C Journal: A Journal of Media and Culture* 8.6. www.media-culture.org.au.

Shuman, Amy (2005): *Other People's Stories: Entitlement Claims and the Critique of Empathy*. Urbana: University of Illinois Press.

Simon, R (2005): *The Touch of the Past: Remembrance, Learning, and Ethics*. New York: Palgrave Macmillan.

Skinner, E (2016): 'Intimate Knowledge: Defining Heritage from the Inside'. In Robertson, I (ed.) *Heritage from Below*. Abingdon: Routledge.

Smith, L (2004): *Archaeological Theory and the Politics of Cultural Heritage*. London: Routledge.

Smith, L, Shackel, P and Campbell, G (eds.) (2011): *Heritage, Labour and the Working Classes*. London: Routledge.

Smith, R (2000): 'Reflections on the Historical Imagination'. *History of the Human Sciences* 13.4: 103–108.

Smith, R (2007): 'R.G. Collingwood's Definition of Historical Knowledge'. *History of European Ideas* 33: 350–371.

Spellman, E (2008): 'Repair and the Scaffold of Memory'. In Steinberg, P and Shields, R (eds.) *What Is a City? New Orleans after Hurricane Katrina*. Athens, GA: University of Georgia Press.

Strangleman, T (2013): '"Smokestack Nostalgia," "Ruin Porn" or Working-Class Obituary: The Role and Meaning of Deindustrial Representation'. *International Labor and Working-Class History* 84: 23–37.

Strathern, M (1996): 'Cutting the Network'. *Journal of the Royal Anthropological Institute* 2.3: 517–535.

Studdert, D and Walkerdine, V (2016): 'Being in Community: Re-Visioning Sociology'. *Sociological Review* 64: 613–621.

Sumartojo, S and Pink, S (2019): *Atmospheres and the Experiential World: Theory and Methods*. London: Routledge.

The Onion (1997): 'U.S. Dept. of Retro Warns: "We May Be Running Out of Past"': https://politics.theonion.com/u-s-dept-of-retro-warns-we-may-be-running-out-of-pas-1819564513.

Thien, D (2005): 'After or Beyond Feeling? A Consideration of Affect and Emotion in Geography'. *Area* 37.4: 450–456.

Thrift, N (2004): 'Intensities of Feeling: Towards a Spatial Politics of Affect'. *Geografiska Annaler* 86: 55–76.

Tindall, G (2006): *The House by the Thames and the People who Lived There*. London: Chatto & Windus.

Tolia-Kelly, D (2004): 'Locating Processes of Identification: Studying the Precipitates of Re-Memory through Artefacts in the British Asian Home'. *Transactions of the Institute of British Geographers* 29, 314–329.

Tolia-Kelly, D (2016): '*Feeling* and *Being* at the (Postcolonial) Museum: Presencing the Affective Politics of "Race" and Culture'. *Sociology* 50.5: 896–912.

Tolia-Kelly, D, Waterton, E, and Watson, S (eds.) (2017): *Heritage, Affect and Emotion: Politics, Practices and Infrastructures*. London: Routledge.

Tosh, J (2010): *The Pursuit of History: Aims, Methods and New Directions in the Study of Modern History*. Harlow: Pearson Education.

Tuan, Y-F (1977): *Space and Place: The Perspective of Experience*. London: Edward Arnold.

Uprichard, E (2012): 'Being Stuck in (Live) Time: The Sticky Sociological Imagination'. *The Sociological Review* 60.S1: 124–138.

Varley, A (2008): 'A Place Like This? Stories of Dementia, Home, and the Self'. *Environment and Planning D: Society and Space* 26: 47–67.

Waterton, E (2005): 'Whose Sense of Place? Reconciling Archaeological Perspectives with Community Values: Cultural Landscapes in England'. *International Journal of Heritage Studies* 11.4: 309–325.

Waterton, E (2014): 'A More-Than-Representational Understanding of Heritage? The "Past" and the Politics of Affect'. *Geography Compass* 8.11: 823–833.

Watson, S (2018): 'Emotional Engagement in Heritage Sites and Museums: Ghosts of the Past and Imagination in the Present'. In S Watson, A Barnes and K Bunning (eds.) *A Museum Studies Approach to Heritage*. Abingdon: Routledge.

Weisman, B (2011): 'Front Yard, Back Yard: Lessons in Neighborhood Archaeology in an Urban Environment'. *Present Pasts* 3: 19–25.

Wetherell, M (2012): *Affect and Emotion: A New Social Science Understanding*. London: Sage.

Wetherell, M, Smith, L, and Campbell, G (2018): 'Introduction: Affective Heritage Practices'. In L Smith, M Wetherell and G Campbell (eds.) *Emotion, Affective Practices, and the Past in the Present*. Abingdon: Routledge.

Whatmore, S (1999): 'Hybrid Geographies: Rethinking the "Human" in Human Geography'. In D Massey, J Allen and P Sarre (eds.) *Human Geography Today*. Cambridge: Polity Press: 22–39.

Woodham, A, King, L, Gloyn, L, Crewe, V, and Blair, F (2017): 'We Are What We Keep: The "Family Archive", Identity and Public/Private Heritage'. *Heritage & Society* 10.3: 203–220.

Woodyer, T and Geoghegan, H (2012): '(Re)enchanting Geography? The Nature of Being Critical and the Character of Critique in Human Geography'. *Progress in Human Geography* 37.2: 195–214.

Woodyer, T and Geoghegan, H (2014): 'Cultural Geography and Enchantment: The Affirmative Constitution of Geographical Research'. *Journal of Cultural Geography* 31.2: 218–229.

Wright, P (2009 [1985]): *On Living in an Old Country*. Oxford: Oxford University Press.

Yarrow, T (2017): 'Where Knowledge Meets: Heritage Expertise at the Intersection of People, Perspective, and Place'. *Journal of the Royal Anthropological Institute*: 95–109.

Yarrow, T (2019): 'How Conservation Matters: Ethnographic Explorations of Historic Building Renovation'. *Journal of Material Culture* 24.1: 3–21.

Young, I M (2005): *On Female Body Experience: 'Throwing Like a Girl' and Other Essays*. Oxford: Oxford University Press.

Part I
Experiencing the past at home

2 Knowing and imagining the past at home

I start with a woman called Yolanda, living in a suburb of north London, who feels a connection to the first woman to live in her 1930s semi. Yolanda imagines the woman in the early morning, in the dark, walking up the stairs to the bedroom with cups of tea – always in the winter time for reasons she doesn't understand, and always as she undertakes this same repeated domestic ritual. Here is a connection through homely practices repeated over the span of years, within the same micro-spaces of the home.

Yolanda also imagines the routes and pathways this woman would have taken as she went about her local chores. When she herself takes a short cut to the shops along a path, her knowledge of its history allows her to imagine the woman taking the same path, which had been the grounds of a private estate, left to the public upon the owner's death a few years previously. The shops might have changed, but the configuration between private domestic and public retail spaces remains the same, and walking the cut-through along the path is an embodied practice of local knowledge which reinforced her sense of this first inhabitant of her home. The path became part of a quietly resonant walking ritual of daily life – knowledge which is given a sensual and emotional meaning through each retelling, each re-ambulation, conjuring the presence of past lives through a sense of temporal continuity.

This chapter explores what gets known about the history of homes. It asks how participants seek and acquire knowledge through particular embodied practices, how they piece together the fragments of what is knowable, how they fill the gaps between knowing and not knowing through imaginative engagements with their homes' pasts, and how emotion is positioned as an extension of, or in opposition to, knowledge.

Researching 'history'

During my travels around the archives and local history centres of England, archivists offered insights into the reasons why people visited. Most came to research their family history, I was told, particularly so in the wake of the long-running BBC television programme *Who Do You Think You Are?* There had been a surge of visitors at the beginning of each series, although the effect was diluting over time. A sizeable minority also sought information about the history of their homes, although many are put off when the hope of effortlessly discovering fascinating stories was replaced by tedious document trails, dead ends and disappointments. As one archivist complained:

> They come and say, 'Have you got the file on my house?' And of course there isn't one [laughs]. They assume it's already waiting for them. ... Some of the television programmes make it look easy, don't they? They only show the results. They don't show the hours of work that's wasted or comes to nothing.

Archivists divided people into two camps: those seeking information about the *social* or *physical* aspects of their home's history. Some people 'got hooked' on accessible census records, whilst others sought specific information: the date the house was built for an insurance claim, details of previous renovation work for a planning application, the location of drains or rights of way. Anna in east London was puzzled to discover that the deeds of her ordinary inner city Victorian terraced house had been signed by a woman from a prominent, wealthy family. Harvey, in central London, wanted to find out why an old wall had been built in an odd place – a mystery never solved. Some people had researched their family history and were seeking a new project, or felt that house history research was a more focused, less daunting, prospect. Anna suggested that researching previous family homes offered a more tangible connection to the genealogical past: 'I can't find out very much about the history of the person. But I could find out about the history of the houses that they lived in'. And for Adam in east London, family and house history was part of a continuum: 'It's a sense of inheritance – being part of something that just grows and grows and grows. But it is all tied up. It's respect for any past'.

Defining 'the past'

What drew people to think about their home's 'past' depended on how they defined what this meant. Laura, living in a flat in a converted brewery, reflected: 'The past I think is all-encompassing. It's funny, you think of it as *one* entity – *The Past*'. Most agreed that there needed to be a degree of distance between now and then in order to achieve the status of 'history' – a word often used as shorthand for something significant, fascinating and to be respected. The yard stick was a gap of at least one generation's span, starting in different periods depending on participants' ages. Laura, a twenty-something, liked the 'vintage' 1950s style, but Brian and Megan, living in an outer London borough, remembered the 1950s well. They didn't consider their 1950s house to be of 'historical' interest because it was within their living memory. For them 'real history' was a time vaguely 'pre-Victorian'. For 30-something Karen, in a north London suburb, the 1950s wasn't old enough either, but she considered her 100-year-old house to have 'history'. Ben, living in south London, sliced the past into different eras, roughly equated to decades marked by 'key moments':

> Part of my fascination with history [is] thinking about the people, you know, in the different eras. You know, so what happened? ... I was a teenager in the 70s. I was a kid in the 60s. I can remember everything about the 60s and obviously prior to that I have an *idea* of the 50s – and then the 40s from my parents. I like to divide time up like that – sort of, it's easier for me. ... I think of the people in this house or another house I've lived in, what they were doing at *key moments*.

Another participant, Nigel, associated his idea of what is old with a house's 'character'. His 1913 house, he said, is 'old enough', it's

> Still got *some* character ... it's still Edwardian. It's all relative ... I guess now you'd still look at 1930s houses as – *slightly* historic. I don't know where you're cut off comes really. I'd say mine still has a more old-fashioned feel than the 1930s houses.

Historical 'expertise'

Whatever the initial impetus, a significant minority of participants had undertaken some form of house research, focused on the building, its inhabitants, or both. A few had published chapters in local history society books, or deposited copies of self-published work in archives, and most had created scrapbooks or folders, usually telling the home's story via a chronological timeline, a house biography. Archivists Chris and Josie, a married couple, had spent 20 years refurbishing their Georgian house in a small market town in the West of England – 'on archivists' salaries'. Their professional knowledge of processes of research and access to resources had yielded extensive information about previous residents. But they seemed relaxed about recording this, deferring the completion of the project:

> We've got a huge falling apart [laughs] file which has never been written up but it's all stuff we've collected on the history of the house ... I mean I've got to write it up in some form because – for my own satisfaction apart from anything else. But just to make the story sort of clear.
>
> (Josie)

In contrast, those with no professional background often took a more serious approach to their research. William, living in a studio apartment in a 1960s council block in west London, had undertaken extensive research on the history of the manor house previously on the site. He had written this up in a beautifully-presented folder. Elsewhere, Brian had also conducted detailed research on his outer suburban London home's place within the locale, and spent hours performing his knowledge during the interview. His wife Megan, who drifted in and out of the room, seemed pleased that he had found an audience for his work. Brian's research was harnessed to a focus on facts and data, his stance one of detachment, as if the history in question was of any building or place of interest rather than his own *home* environment.

For participants with more direct experience of public heritage or conservation projects, their view of the history of their homes – usually focused on its *physical* aspects – was also consciously rendered 'unsentimental', 'hard-headed' or 'realistic', reiterating an assumption that rigorous research does not, or ought not to, involve the emotions (Weisman 2011; Berger 2007). Heritage consultants Derek and Emily, living in an outer London suburb, described the details of their home's history at great length. Derek commented: 'I think I've done too much work on various historical projects to have a sentimental view of the past'.

Elsewhere, Jacob, living with wife Penny in a market town in southern England, had a similar response. As treasurer of a county archaeological society, and a previous university employee, he 'had to do long-term maintenance plans, septennial reviews of historic buildings'. This led him to view old buildings in a particular light:

> I don't think it's a kind of romanticism about, 'Oh, people used to live here in the past', as it were. No. I'm too much of a historian to kind of be *sentimental* in that way. Penny may offer you a different view. She is a non-historian.

Being 'too much of a historian' to be sentimental suggests a belief that professional knowledge creates or requires emotional or imaginative *distance*. It led Jacob to view old buildings in a 'fairly hard-headed way'. He implied that his 'non-historian' wife might be more romantic or sentimental, whereas if you have an expert's eye (Yarrow 2019), knowledge of a house's history is there to be revealed: he knew and respected 'building experts, building historians, you know, who can deconstruct a building – work out the sequence in which things have been done'. Working out the 'sequence in which things have been done' in the past, Jacob added, was more important to him than imagining the home's past residents. It enabled him to 'think about how – was the house *organised*', informing decisions made during renovation work.

Archivists Chris and Josie were able to use their knowledge of the social history of their home to construct a genealogical timeline of past occupation, matching this to the age of the house's features. This allowed them to work out which families did what renovation: 'You just think, "I know which family did that"' (Josie). They also contrasted their own engagement with the past with a more fanciful response, as Chris reflected:

> Not, sort of, with any – *emotional* attachments about, oh you know, 'this has been touched by a dead person', or, you know, 'you don't want germs that might be on this from 200 years ago'. I mean I think it's just a more practical, hard-headed point of view that, you know, having worked in

the sort of jobs we do, just transferring it to a house. It's the same sort of thing – attitude – really, you know. It's still handling old stuff – and looking at it from a *practical* point of view. You know, at work – 'What do the documents tell us? What use can we put them to?'

Despite this, Chris was aware of contradictory responses: he *did* relate to his home's 'atmosphere' and imagine past lives in a more immediate way:

> We want to be surrounded by [the house's] atmosphere, its historical asso-ciations. ... I almost sort of contradict myself because I said I like to sort of imagine people sat here say, you know – 100, 200 years ago, doing the same thing that I'm doing ... I just like the sense of continuity.

Embodied knowledge

Most participants did not carry out extensive research into their home's past, their understanding emerging rather through a mixture of chance encounters with knowledgeable others and through visceral home-making activities such as renovating rooms or digging in the garden. Here discoveries from the past were not sought or planned but part of a spontaneous, usually pleas-urable, process of stumbling upon and revealing the home's traces – the home momentarily becoming an accidental archaeological site. Even Jacob conceded this process involved feelings of 'fascination': 'The challenge of his-torical research – some people do jigsaws ... There's the fascination of the chase, you know. Some people – I don't know, hunt foxes. There is a goal. You want to get there'.

Bella and Sylvia, a mother and daughter living in a large historic manor house near Bristol, knew this perhaps more than many given the age of their home – a Norman hall extended incrementally over centuries. The family, who had lived in the house since 1922, described numerous chance 'finds' hidden behind layers of plaster and wallpaper, as Sylvia recalled:

> We'd had to do a bit of mending. We were *finding things* which we hadn't deliberately set out to do ... The plaster was coming off and we were going to have to replaster it. And when we started to do it, we found things underneath. It was just chance. ... It is interesting, isn't it, what we've uncovered? By *mistake*.

One discovery was made by the family dog:

BELLA: We know there's a staircase behind there because Georgie [the dog] found it
SYLVIA: They filled it up with stones ... We actually had to do it because the squirrels were getting in
BELLA: She found a squirrel's nest underneath the wood

Elsewhere, in an early Victorian house in south London, Leah described the excitement of finding a hidden fireplace:

> We uncovered a *fireplace*. Because (ex-husband) Alex said: 'Oh there is just this bit of wood' – in front of a fireplace like that, and it's sort of painted. And he goes, 'Oh there is no point taking that off, there will be nothing under there'. I was like: 'Get it off! Get it off completely!' Cast-iron, original fireplace underneath it. Spent ages paint stripping all the stuff off ... *Amazing*! It felt like I had uncovered a mummy from the pyramids or something ... It was completely intact, you know. It's like an archaeological find.

The excitement was not just that these things were hiding – and the joy of unexpected discovery – but that the fireplace was *intact* and *original*; the inheritance of something pristine and unsullied by time.

Beyond the discovery of hidden fixtures, knowledge of the home's pasts was revealed through renovation work, often involving stripping back the home's surface and structural layers, adding layers back. This hands-on process became a form of emergent embodied knowledge of the home's physicality, increasing people's emotional relationship with it. Pam, who renovated a ruined farmhouse in an isolated hamlet deep in the Yorkshire Dales, described this as

> an extra bond – if you do physical work in a house. If you contribute to the structure of the house. For me, anyway, there was an extra connection. So that probably explains why I feel so strongly about this house.

Archivist Josie also described her 'intimate relationship with the house' emerging from knowing 'not *quite* every inch', to which Chris interjected:

> Oh I think I probably do. Every inch of the house. Because I've been under every floorboard, into every corner of the roof and – yes. I was scraping paint so I've got to know this bit of wood very well, you know [laughs]. ... I do know every inch of this house, you know. Hands on and *everything has a story behind it*, you know. There's – oh – the doorknobs that – I remember I bought these, you know, second hand. And the fingerplates. And, you know, the little *bits*, sort of architectural fittings and things.

Getting to know the house better through renovation work is also about adding to its 'story', forensically revealed through the laying bare of its materialites. Here the home is not a static, immobile container to be observed. Knowledge of a home changes people's intimate relationship to it, just as the home itself is changed through renovation work. Through such a process, new stories and memories are added – including what has been changed by participants over their own time in the home, as Chris said:

I know the story behind so many bits of the house that it's going to make it really hard to lose this house because it's all part of my life, you know. I can remember when I did this job and when I did that.

Elsewhere, in a garden square of early Victorian houses in east London, married couple Martin and Carol lived in a bottom flat, but had later also bought the top flat, which had needed renovating. Martin described his intimate connection to the 'bones' of the house through this process:

> The place upstairs was a nightmare to gut. But it made me much closer to the actual building itself. ... If you like, the house became much more part of me ... I then began falling in love with the house itself.

The home revealed itself through these haptic, visceral rather than detached forms of encounter – in some cases, the past *literally* getting onto or under the skin. Martin recalled:

> One of my sons was up there chipping away at the plasterboard, and the whole roof caved in. So I went up there and I found him. He was half naked anyway because it was just hot and sweaty – but all I could see was this black powder with – two eyes ... If you think about all the pollution around this area in particular because it was quite an industrial area. *Soot.* Soot *that* thick had just obviously gone in through the tiles and was stopped by the plaster that was on the ceiling ... The whole thing collapsed. So there he was with these great big white eyes [laughs] – just black head to foot.

This comedic narrative, vividly capturing the shock of the ceiling's sudden collapse, was extended to a reflection, through the visceral encounter, of the area's industrial past – the soot coming in from outside, through the cracks, 'through the tiles', and held in place by weakening plaster, waiting to be released.

Harvey and Rochelle, living in a Georgian house in central London, told a similar story – the messy inheritance from the area's industrial past also defying modern cleaning appliances:

ROCHELLE: Harvey went up there on New Year's Day. In the loft was – piles and piles of *soot* ... It was part of the history of living near three main-line rail stations ... My grandfather was a coal miner. That's why I wanted Harvey to wear a mask [laughs]

HARVEY: I took three showers to get the soot off and woke up with soot on the pillow ... It was like a winter wonderland, you know – these little billows of, of black smoke ... We had had the roof done and all the broken bits of slate had fallen down into the soot. At first we got an industrial Hoover to try to suck it up. But you would suck for two seconds and you'd get a piece of slate. And finally there was nothing else to do but to go down with a dustpan and brush between the joists, one at a time … You know those big blue bags used by builders? I filled up 22 of those.

Previous uses of spaces

Beyond spontaneous embodied knowledge of the home – emerging through messy bodily encounters and the hidden discoveries of the intimate taking away and giving back process of renovation – participants also attempted to trace earlier configurations of the home's layout – of rooms, corridors, staircases, windows, doors and fireplaces. Here, the home's materiality revealed how past residents had lived – their social worlds mapped onto the changing walls. The material spaces and fixtures acted as both historical evidence and triggers for the imagination. On guided tours, participants showed what they had gleaned or what eluded them of the previous layout, often reconstructing this through fragments, marks and other material clues which allowed them to 'place' previous residents within the home. These traces of the past could be intimately suggestive – ghostly lines or small indents on walls where doors or windows might have previously been, or faint chalk marks on outside walls where children had recently played.

Some homes had changed more than others. In a small town on the Thames, Gemma, a piano tuner, described the evolution of her unusual cottage, only accessible down a long narrow pathway. It had started as a well (now under the living room floor) and then became a Victorian laundry. Later it was extended into a cottage.

Some participants followed up specific clues – such as researching when saw mills started making floorboards, to date the floorboards by their width and symmetry, or observing that the locks on the doors of a home might be a sign it previously had been separated into flats. Rochelle and Harvey in central London speculated that their house had been a brothel based on information they received about the street prior to its recent gentrification. They pointed out what they had considered strange – the 'awful lot of plumbing' in the upper rooms, assuming this counted as evidence.

Imagining past residents often focused on particular spaces and times – moments in the calendar year, the seasons, or local impacts of national events. Ben in south London wanted to know what past residents did, 'you know, on Christmas Day, *special* occasions', but also the details of their lives within the home. He asked:

So what would they have been doing on a – summer's day? Would they have enjoyed the garden, or not enjoyed the garden? You know. Who planted the trees? There are quite a few old trees in there. So, they are very much part of the house … What were they doing when the bombs dropped? What were they doing, you know – of an evening? Would they go down the local pub, or not? What games did they play?

Participants also discovered details of specific past residents that enabled speculation about their presence in particular rooms of their home. In an Edwardian house in south London, Susan had been lucky enough to be given a photograph by the granddaughter of one of the first inhabitants, a 'Victorian lady' who took to her bed with an unspecified illness. The photograph shows the woman sitting up in her bed with a book on her lap. Susan realised that the bed was in the same place in the same room as her own bed. It also allowed her to speculate about how the woman communicated her needs:

> Interesting that in this photo she's got a bell for the servants. Can you see? There's a servant's bell. It sort of hangs down there … They've got some sort of – *thing* coming off of it. Like a wire she can pull. So she's sitting here, and she pulls it. But it's not that big a house. I suppose if they are in the kitchen.

Having worked out where the kitchen used to be, she suggested it was too far for her to call to summon help.

Elsewhere, when Irish photographer Julia moved into her terraced cottage in north London, her neighbour was keen to tell her that the poet John Keats had lived there for a short time. She speculated that he might have rented the living rooms:

> [He would have slept] here in this room. Yeah, almost certainly. These houses are designed so that the living area is upstairs – I don't know for *sure*, right? And the kitchen was downstairs and so the landlady – and there *was* a landlady when he lived here – probably lived downstairs near the kitchen, right? And rented the upstairs rooms. So one of these would be a bedroom, the other would be where he wrote and ate and washed, and whatever. I'm pretty sure that he would have been up here. I mean, I might be wrong of course, we can never – we can't *know*, there isn't any *evidence*, right, unfortunately, except to say that there was a landlady and on the practical basis it's likely that he was up here.

In turn, after reading that he had been seriously ill during the few months he spent in the house, she looked at the grate in the front room with a new understanding:

> I think the grate is original. It's a bloody awkward grate actually. Keats would have found it really hard keeping this room warm with that grate, right, because it's very shallow – it's not *deep*. So it's very hard to – get a fire going in it, right?

As will be explored, these more permanent features can trigger particularly intimate images of past lives. Pam in her Yorkshire farmhouse showed an original stone staircase, indented with the imprint of wear by successive generations of inhabitants. The stone was shiny from water erosion, signalling a symbiotic relationship between the home and its landscape: 'There are stones that have been shined. And you can see that they've taken it out of the [adjacent] beck, because this is water-eroded stone. Those steps there'.

William, in his south London council flat, had spent time in prison and recalled his 'home' in an old prison cell. He discovered this dated to 1790, still having the original door and walls (with 'lots of gloss paint on') and, as he speculated, the same 'lack of light'. These unchanging characteristics triggered reflections on the lives of the early inmates, a compensating fantasy mixing empathy for their lives with reflection on his, as he saw it, less harsh circumstances:

> As I lay there I thought, 'Alright, so I've got smelly socks next to me there' – which I couldn't *bear*. The two blokes were alright. … But *imagine* if it had been the old days. I don't know what it would have been like then. … That would have been built for … Napoleonic? They were for soldiers who had had breakdowns.

The limits of historical sources

For some participants, no such material clues were available to help them understand their home's past configurations. One complained:

> And it's kind of like working out how it would have been done, and we still haven't been able to work out how they did it … My research in 1911 shows two households, but it doesn't say where they *were* in the house.

Jacob in the southern England market town had a similar complaint: 'I'm interested in the situation of housing immediately post-war. Was this subdivided into bedsits or what? We don't know. We don't *know* … The electoral register just has a string of *names*'.

For archivists Chris and Josie, faint traces on walls had offered tantalising clues, but these only served to remind them of what they couldn't know. Producing a photograph of the front living room as it had looked when they moved in – in a raw state, 'untouched by the 20th century' – they pointed out the walls. These contained numerous ghostly outlines of various shapes:

> You could see spaces on the wall where pictures had been for years and years and years obviously because they leave a paler space. I always wondered what it looked like. I would have liked to have seen it. … Some of them would have been portraits or photographs of people, you know. … And I just would have liked to have *known* what was there before.
>
> (Josie)

Not being able to *fully* know or visualise the past of the home as it materially looked or was decorated by earlier residents is a source of great frustration.

This is reflected in fantasies of time travel. In the manor house near Bristol, Bella and Sylvia expressed this desire – to glimpse the past *in* the past:

BELLA: You want to be able to go back in time. Just for a second or two. To have a look. I'd *love* to see what it was like. Because I want to *know* … Because it's just *fascinating.* Why did they do that? [Whispers] 'Why did they do *that?*' What made people put that *there?* … I would like to see what the house looked like when they had finished the Jacobean project [wing]. I would like to know what it was like before they *started* it.

SYLVIA: Because we think it probably was falling down, that bit … We don't know. I call it semi-educated guessing [laughs]. One just doesn't know.

For other participants, the desire was to glimpse earlier *inhabitants* – to eavesdrop or even *enter into conversation* with them. Penny in the southern market town imagined herself 'walking through the front door' in an earlier time:

I try to imagine what it was like being inside the heads of the people who lived here. And also – how it would have been lovely to … in some way or another, to have *visited* them, *spoken* to them … I just really wish that I could just kind of wind back and – walk through the front door and see what the people at that time, whoever they are – whether they are the first inhabitants or the – later Victorians and Edwardians. What they're wearing, what they're talking about, what they're eating, what their *worries* are.

The 'little things'

Penny's desire to know of past lives' clothes, conversation, food – and *feelings* – reflected participants' interest in the 'small stories' (Cameron 2012; Webber and Mullen 2011; Shuman 2005). Karen, in her Victorian north London terrace, recalled helping to redecorate her parents' house and discovering that 'people hadn't bothered to take the wallpaper off, so you could kind of strip down through history. It is the *little things* like that which fascinate me'. For Gemma in the town by the Thames, the focus was also on what she called the 'minutiae', which related to her concern for the social history of ordinary people:

I've always been interested in history. Not so much *studying* history – the repeal of the Corn Laws and all that sort of thing – but *life*, in the past. What I call *'real life'* … I just love the *minutiae* of everyday life, but from years ago.

A majority of participants shared this fascination with the 'ordinary' and unknown, reiterating ideas about the democratising value of family research (Cannell 2011; Robertson 2012). As Yolanda reflected about her 1930s north

London suburban semi: 'It might just be an ordinary little house in an ordinary little road, but it still has a *history*'. This focus on the 'ordinary' pasts can also offer a means of reflecting on people's *own time* in the house. Australian-born Anna in east London, for example, described her desire to create a scrap-book of the house's history which would focus on 'just little things about what happened and where and why'. It would contain not just the history of past residents, but a mixture of personal and more public histories and the various selves and others she has encountered or imagined. Reflecting on these inter-woven histories was a further way to differentiate between 'expert' engagements with building history and the home's past as something containing personal meanings for people when such buildings are their *home*.

Rita had inherited a large manor house in the Sussex countryside from her father, who in turn had been passed it by his brother, an artist who had died childless. For practical and financial reasons, she had been forced to sell the house, buying an ex-council flat with the proceeds. But she continued to be nostalgic about the house, remembering her 'magical' childhood exploring it. She contrasted these memories with the tone of a report written by a histor-ical expert:

> There was a guy who was linked with Preservation of Rural England. He said, 'Can I come and have a look?' And I mean he was a kind of nerdy type, really interested in the alterations. He'd say, 'Yeah, you could see where that's been done, and that's been done. And that's where a lean-to was added' … It's quite technical [reading his report], 'A lesser two-storied rearranged with walls of flint … and less contemporary with the Period B works. Another 18th century alteration was the rebuilding of the Period B ground floor walls and brick', etc. etc. You know. It was – quite *dry*.

She argued that such a professional attitude, with its abridged, technical lan-guage, offered a limited understanding of the home's 'complex' history – the memories of its quirky spaces, atmospheres and smells:

> There's so many aspects of [the house]. You know, from the workmen that rebuilt the interior walls, to my father's memories, to *our* memories as kids. You know, all the rooms had names – the Long Room, the Pipe Room … the Dressing Room. And then the bathroom was truly amazing, with a floor that really was very, very slopey … dark brown polish sloping away from the bath. The house has got so many aspects to it. … It had this sort of aura, you see … a kind of a smell and atmosphere and really old furniture. … It was a lovely smell as soon as you went in. … I don't know what made it smell like that. … It had its own kind of history. It's a complex thing.

This 'own kind of history' suggested a form of affective heritage of the home, of sensual memories of its spaces and smells and aura – the *feel* of the place.

Imagining domestic routines in situ

For most participants, personal memories gave way to imagined scenarios of past lives. In some cases, the past needed little exertion of imagination or time travel fantasy – the photo of the Victorian lady reading in bed with the little bell, the shallow fire grate suggesting the frail Keats' bodily discomfort in the house. For many, there is less tangible evidence available, but people could try to imagine previous residents in situ, how they might have *fitted* their lives into their homes. Susan, for example, had a particular interest in a woman, who she discovered had built her house during the early twentieth century. The records showed her to be a seamstress, fuelling curiosity as to how she had managed to acquire the land and instruct the builders. She reflected how 'I only seem to be interested in her as it pertains to this house', adding that this narrow interest related to all the residents:

> I haven't tried to find out what happened, you know, where they came from and where they went. I haven't tried to branch out – I only seem to be interested in them in connection to this house. I haven't tried to find out what brought them here or sent them away.

In turn, interest in imagining past residents in the domestic interior tended to focus on their domestic practices – taking tea up the stairs to the bedroom, sitting around the fireplace, playing games or walking out into the garden. Gemma's fascination for a late Victorian laundress led her to visualise her living and working in her cottage:

> She lived here too. From what I can gather, that was a little living area. … But I think she ran it as a laundry herself, off her own bat. … And her name was Liza Brooks. … Because this has high ceilings, I just imagine there being lots of clothes hung up drying and the fire always on the go under the copper. … Somehow I always see her sleeping in *there* but I don't know why.

Gemma also described seeing and sensing 'Liza' as a ghost, but one 'confined' to the rooms of the building before it was extended into a cottage: 'I have never felt her in the music room or our bedroom, which would have been the garden in her day. So she seems to be confined to *her* house'. In a similar way, Gemma's imagination confined Liza to where she worked and slept in the cottage.

Yolanda's fascination for the first woman of her 1930s suburban home is also narrowly focused, as she reflected:

> It's only the domestic bits that I can imagine, isn't it? I can only imagine her placing the breakfast, using the space. They probably would have used the dining room which we actually don't use very often because our kitchen is extended so we tend to just use that. The kitchen was half the size.

This focus on the woman within the home also reflected her life as a house-wife and her daily routines. As a woman, Yolanda argued, she wouldn't have had any other life anyway. Thus the *placing* of her within the home reflected her understanding of her role as a suburban wife and mother during the early twentieth century:

> I think about her life. And – yes, about her life. Not really about her as a person, [pause, quiet chuckle]. Well – women weren't really allowed to be very individual in those days, were they? They had to – they were housewives, they were mothers and housewives. I imagine she didn't have a career, and I imagine that's what she did *all of her life*. I might be com-pletely wrong, but that's what I imagine.

Given this, she drew on her local knowledge to focus on what she would have been *doing* within the home – reflecting her time and practices within the various spaces of the home and environs, rather than dwelling on her as having an individual personality:

> What was she doing? There are a few little shops down the road. So she might have ... walked across the fields, or she might have walked past the beautiful big houses with her shopping bag and picked up her fresh butter and cheese from Sainsbury's. ... If the children went to the school down the road, they would probably have walked down the road and through the park ... past some allotments, some shops. There was a greengrocers there when we arrived, long gone. And there may have been a butcher, a baker.

The 'knowability' of past lives

This focus on reflecting past lives in their local context, and the embodied routines of domestic life, further reinforced interest in inconspicuous or 'small' aspects of everyday life. But it also suggested the limits of *knowing* past residents beyond their generic aspects. Whereas for those such as Penny, there is a desire to understand their inner emotional lives ('what their *worries* are'), for most there is resignation that such a level of intimacy is not possible. As Yolanda put it, even without the distance of time it is difficult to know what happens:

> People's lives – you never really *know* anyway, do you, what goes on behind closed doors? You'll never know what another family's life is like – unless you live with them, not just spend maybe a week or two, you actually have to *live* I think with other people to find out what their lives are like.

Other participants reflected on the inability to know past inhabitants *as individuals*. Pam in Yorkshire described how for her imagining past residents does not mean making 'people' out of them. She didn't conjure them beyond being stereotypes or just the people who shared the same home. She wondered why she didn't think about their 'stories' given her love of reading fiction:

> I don't try and create a person out of them. They are relatively faceless. I don't know why … I read a lot. I read a lot of fiction and there were stories there, and I haven't thought about *their* stories, what their stories might have been.

Lillian, living in an old house on a major arterial route out of south London, described imagining the 'costumes' that people wore in the past – in her focus, tricorne hats, breaches, and long waistcoats – rather than reflecting on the people themselves: 'If I hear something say that happened in the 1770s I think, "Wow, people like that were walking through my house in that costume", and everything'.

But even where there is an 'actual' face available, staring out of the past into the present through photographs, people seemed reluctant to assume they can 'read' them. Susan received further photographs of the Victorian lady in her home, including those of her posing in her bath chair in the back garden. She reflected on the woman's posture, expression and demeanour, but couldn't go beyond a generic engagement:

> I think they are a bit one dimensional actually. I do think of her as this funny sort of Victorian lady. There she is *there* – and she just seems to be terribly one dimensional and stiff-backed, a formal Victorian lady.

Looking further, a more enigmatic relationship between individual and era emerged:

> Normally you would be quite posed. She's not smiling, is she? Maybe she didn't smile. Victorians – *didn't* smile perhaps. She looks as if she's – *kind of* smiling in that one … I think that's her – *there*. You can just see, you can *just see* a half-smile there. But there she's *very* serious. … Look, even in the deckchair she's sitting upright. And even when she's leaning – look, that's her there – she's a bit stiff [laughter]. Maybe she's still got her corset on.

Even when there appeared to be an abundance of historical information and detailed commentary about a past resident, there is still anxiety about assuming that past lives can be fully 'known'. Julia had become fascinated by Keats' life after discovering his connection with her home. But after extensive research, she concluded:

> I don't think you can really *know*. You can have a fairly good idea about some aspects which were commented on by people. But then you have to take

into account their personalities and their pre-occupations and some of those you – *won't* know. So you can't take anything for granted really, you know.

Census names

Responses to census records – where the households, their names, ages, occupations and relationships to each other are laid bare – provided further possibilities for reflecting on past residents as individuals. It was the *names* of residents which garnered the strongest reaction. Certainly, some participants' engagement with 'the past' focused exclusively on a particular era. Abigail, for example, was only interested in the original occupants of her Georgian flat in north London – 'all the rest are just names' she said dismissively. Jacob in the southern market town parodied any imaginative connecting with previous residents assumed to be possible through knowing their names:

> I don't think I am thinking: 'Oh, Dr Smythe' or whoever it was, you know, 'would *recognise* this room'. And – or if he'd been here in-between time, say: 'Hasn't, you know, Mr Lambe messed things up?

But for others, the names reminded them that past residents were *individuals*, and this might help to 'build a picture'. Carol, in the east London square, reflected:

> It makes me realise ... that this shell has contained lots of other different lives with other different stories. To me it's *the stories* of the people that have – it gives them a *personal* – it gives me a sense of them *as people* as opposed to the fact that you know lots of people have lived in this house before. Now I actually *know* what those – who those people *were*.

The individual names bridge the gap between generic facts ('lots of people') and personal 'stories' offering a 'sense of them as people'. The latter is a different kind of knowing, a shift from 'what' to '*who* those people were'. This in turn created an uncanny frisson for some. Adam in east London described the handwriting of the 'head of household' in the 1911 census as 'eerie' ('Got lovely handwriting, hasn't he? Really gorgeous. It's kind of eerie, isn't it, as well'), whilst Karen in her north London terrace also reflected that looking at the names on census returns is 'weird' because it made the people feel more 'real':

> It's weird to see the names – to make them *real* ... They've got names and ages and – they are real people not just – you know – with jobs and – yeah, it makes me interested to kind of find out more. You know, *why* was his sister living with them? [Laughs].

This tended to be a minority view. For most, the names tantalised but, like the outlines of the paintings on the walls of Chris and Josie's house, they can only ever hint at past lives. Historical 'data' is limited. It doesn't allow them to *know* anything or anyone on a personal level; they remained past 'others'. Leah in south London described how her initial interest in the history was focused on specific issues such as what people did for a living and how many children they had. But she added:

> And then I tried to imagine in my head what it would be like when they were living here. But then you can't – I realised actually the *limitations* of the census. Like, when you get this knowledge and you find out the names and everything else, you think, 'But that's *all*'. I was a bit frustrated because in a way I thought, 'That's all I can ever know'. [Shuffles census papers on table] *this* doesn't really tell you – it just gives you *names*, doesn't really tell you *anything* really. And when people look at *our* stuff they'll just see *our* names. It's just *data*. It's just kind of *raw data* … It doesn't go deeper, does it? It never will.

Julia made the same point: the census records offer a mere 'outline' of the past: 'The information is so scant. I mean it's interesting to know what they were doing, right … [But] it's like a diagram really, compared to a painting, isn't it? A bare outline'. This made it hard for her to find any 'meaning' in the data, to offer enough information to allow her to 'think about' past residents:

> It *doesn't mean* much to me. It's hard to think about them. You know. It's hard to know *what* to think about them. It's not that they're not interesting, but it's just difficult to know – how to make a connection with them, really.

But elsewhere, Sara, living alone in a maisonette in west London, initially had a different response as she scanned the census records: 'All these names now that I had no knowledge of whatsoever. … I think a name does add something. … It does personalise them'. She then attempted to make a connection by reflecting on her sense of past residents as *similar* to her:

> Obviously human nature doesn't change that much, so it wasn't all completely different. The situation, the circumstances, were different. But, yes, yes. They had the same sorts of feelings and aspirations and worries, and all the rest of it. Just in different, a different context really.

But she then added that the records only offered fragments:

> You are just speculating because inevitably you've got this information, you're trying to interpret it. You've got these little snippets about their lives … It does open up so many possibilities, doesn't it? About what was going on in the lives of really quite a lot of people. But it is only one bit of information, isn't it? It is very sparse really.

The exercise of the imagination

Awareness of the gaps in knowing past lives in the home led some participants to reflect on the role of imagination as the only recourse to creating a connection with past people – one which could never be 'real'. Miriam, living alone in a council block in central London, admitted how the desire to imagine did not lend itself to an accurate picture of the past:

> It's not making connections with *real* people at all. It's just imagining – some *past* I suppose. It's *my* version of the past. I suppose if I was actually there it would be quite different [laughs] ... It's a romantic interest, really. Thinking about the past and who might have been here. What the neighbourhood would have been like ... When you look at the Georgian squares down there, you can imagine them being people's *homes* and people coming along in a carriage, coming home by gas light ... [It's] about not being very scholarly or academic about it. Just imagining perhaps a scene that might have been here. Different furniture and different décor, and people with different interests – perhaps not quite so many books [laughs] ... Your imagination might include things which never actually really happened.

In a similar way, Samantha, living in a suburban flat with her husband and young children, described how reflecting on the past is always an 'exercise of imagination', a conjuring of metaphorical ghosts:

> It's about, I guess, the exercise of imagination, you know, thinking alternative narratives that have also happened here. And, you know, the kind of, the *ghosts* of – I don't actually mean *literally* ghosts, but metaphorically, yeah, that kind of ability to imagine what happened.

But for Yolanda, this ability to imagine is more pleasurable than the actual *knowing*; she enjoys the 'mystery' of the past:

> I think being able to *imagine* things – as they were – there's, there's some degree of pleasure to be, to be got from that ... There has got to be the mystery and there's nothing I can do about that at all. And actually I quite *like* that.

Whereas for these participants the acknowledgement of an inability to *know* was expressed as varying degrees of frustration or enjoyment, for others there was relief in the ability to distance past lives from their own. Anna in east London, who did not like the people she bought her house off, expressed her belief that imagining past lives offered more leeway than actually knowing them: 'They might have been horrible. But at least ... you can just imagine what you *like*'. Ben in south London described how he was 'interested about how they lived their lives and who they were', but that they remained distanced: 'I don't see them as people yet'. In any case, he added, actually knowing them might not be so enriching: 'I'm sure we'd have nothing in common'.

Embellishing and fabricating stories

Ben added that the inability to know previous lives led to a fictionalisation of the past, with his 'imagination running riot', in a form of 'play ... like writing your own novel, really'. This was a danger – trying too hard to speculate about the past, to fit together the fragments of scant available knowledge, could lead to fabrication – the imagination here in opposition to knowing. Sara in west London worried about this:

> You're then applying things you know and ... possibly putting together a bit of speculation, but you have no idea whether you're on the right track or not. So you have to in a way try and not go too far down that route, because you could be making up all sorts of – you know, *stories* about them that could be completely untrue.

For Leah in south London, such a fantasy was far better than reality, offering a gloss on the past which allowed a degree of control in recreating it as she wished it to have been. With knowing irony, she described how worrying too much about the truth would 'destroy the glamour':

> I suppose I don't really think about all that business about the *real people*. Because that would destroy the glamour a bit. Because if I actually knew what they were like, if I actually met them and they were like some normal family, that would destroy the fantasy. And the fantasy for me – I just want to see them as 'Wow, these Victorian type people' [laughs] ... In a kind of childlike way, in a kind of costume drama like way. I don't want to imagine them as just like – *us* ... If the Smiths suddenly appeared around the table, you'd probably be disappointed. You'd think, 'Oh'.

Like others, Leah didn't want to know what past residents were like in case they turned out to be ordinary, dull. She wanted the past to glamorise people, enjoying this aspect of distance from their 'reality'. She preferred them to be one-dimensional generic beings in old-fashioned 'costume drama' clothes – safely exotic and not *here*. But she mused that perhaps such a position was also a reaction to her frustration at not being able to know very much anyway:

> That's why I had to romanticise it as well, to satisfy that bit of frustration. So that's why I turned it into, 'Oh yes. All these people with bonnets swishing around, black widow type dresses on'. And things like that [laughs].

For others, not knowing the past allowed scope for playful forms of imaginative engagement to be taken further. Samantha and her partner Jake had bought their flat from a man who worked in architectural salvage. He enjoyed telling them stories of the objects he had placed in the flat: light fittings from the Champagne Bar of Royal Ascot, the bathroom cabinet made from an old fire hose cabinet from the House of Lords, marble panelling in the bathroom salvaged from a bank. But they doubted all his stories, as Samantha said: 'I don't really know how accurate he was … It was a bit sketchy … I wouldn't absolutely rely on his account'. On the other hand, they had entered into the spirit – of not letting a good story get in the way of the truth – creating their own 'myth' about the marble panelling. They told themselves that this had been salvaged from the European Bank for Reconstruction and Development, the subject of an infamous overspend scandal which led to the resignation of its president; the lavish marble lobby in its new headquarters had become emblematic of management excess. Samantha admitted that they liked this idea but it was 'pure myth-making' on their part: 'There are plenty of other banks'. Her Irish family had also entered into the same spirit. Clearing out an old family home in Ireland following a death, she had inherited a 'really uninteresting looking table' that nobody else had wanted. On the underside of the table was a label explaining it had been built using 'oak from the roof of Westminster Hall' when repaired in 1934, the oak in turn from a forest which was '1600 years old'. Her family, having heard the story of the cupboards allegedly from the House of Lords, claimed it was highly 'appropriate' that Samantha should have the table from Westminster Hall – pointing to the so-called coincidence of these family and house inheritances. Samantha joked that this had been their ruse to 'get rid of all the stuff'. The story of the table, she added, was in fact also highly dubious – perhaps the only real connection it had with the bathroom cabinets from 'the House of Lords'.

In his east London house, Adam had also been told a story – that a famous music hall star, Harry, had been a former resident. This had caused him great excitement, at the least it 'made life more interesting'. He added: 'That's real fame, isn't it? That's *real* fame'. Unfortunately, the census records *did* list a music hall performer called Harry, but a man with a different surname:

> So whether somebody knew it was a musical artist whose name was Harry and just spun the fact that it was Harry, do you know what I mean? That would be – I would be *devastated* if that was true. It might be true. I'm quite willing for that to be true … Anybody that lived in this house would be of interest to me. I'm fascinated by what they did and how they lived, and how many people were here.

The focus shifted defensively from disappointment at the possibility that he was wrong about the famous resident to professing a more egalitarian interest in *all* previous residents. Adam's wife, briefly entering the conversation, commented that Harry was a 'nice story ... a bit romantic', but they were 'not sure if it was true', to which Adam retorted: 'I've said that. I have *confessed* to that'. As with Leah, there was a knowing line between fact and imagination. They wished to persist in the fantasy that Harry lived in their home; even knowing that it might not be true, they held on to the idea because it was 'romantic'. Adam added that in any case, *heritage* officials at historic sites always embellished the truth when telling anecdotes on tours, so why couldn't he? 'I'd make it up if I had to. I would. I would make it up ... to make life more *interesting* for me'.

Figure 2.1 Outlines on walls show where pictures would have hung.

Conclusion

This chapter has explored the ways people define and give value to different forms of knowledge and imagination in coming to terms with the past of their homes. Whilst some participants had carried out research into aspects of their home's history – and were invested in an 'unsentimental' notion of themselves in relation to such a process – for most a more embodied and often viscerally intimate 'knowing' developed spontaneously through chance finds, or emerged through home-making activities, through the act of living and taking care of their homes. There are overlaps between experts' methodical unpacking of the changing layers of a building's physical biography, and participants' own attempts to piece together past configurations through the perceptible traces and fragments left behind. But whilst some participants 'performed' their knowledge gained through archival work, or wished to reinforce a belief in the value of a distanced stance in pursuit of historical 'truth', most described their engagement with the past as emotional and imaginative, fuelled by a fascination and desire to connect with their homes and its various histories, if sometimes frustrated by the limits of such a project.

Knowledge and imagination are thus often presented as an oppositional binary – the former sometimes given higher status or, alternatively, assumed to be too narrowly conceived by emotionally-distant 'experts' (Berger 2007). In a few cases, there appeared to be a gendered element to this, with husbands assuming their wives to be more 'sentimental' due to their lack of historical training (Blackman *et al.* 2008). But in many of these accounts relationships between knowing and imagining the past emerged as part of a more creative exploration of responses – if at times an ambivalent one – as participants grappled with what they can or cannot (as well as what they wished or did not wish) to *know*. In most cases participants *themselves* were critically aware of the possibilities and limitations of their different encounters with the past and of the forms of 'historical imagination' (Harvey 2015) they employed. They were thoughtfully reflective of the reasons they may be drawn to fill the gaps of what is knowable through embellishment, conjecture, compartmentalisation or sugar-coating, and how these responses might lead, indeed, to a 'falsification' of the past. And, for some, there was a need to maintain imaginative 'distance' between *now* and *then* – less as a technique of rigorous historical enquiry and more as an emotional reaction to the sense of visceral nearness of the past at home.

All these responses suggest we need to engage with how forms of everyday heritage are practiced and experienced in a range of different ways and through a wide sensual and imaginative palette, and how forms of historical 'knowing' – of pasts one has not lived through – are experienced in everyday contexts and emerge out of the micro-spaces of the home. This focus allows insights into engagements which might value different forms and practices of knowing or experiencing above others. It offers perspective for understanding how particular forms of historical enquiry and encounter are valued, beyond the expert-amateur binary (Schofield 2014). Central here is the fact that participants *themselves* are aware of the relative limitations and possibilities of different approaches to understanding the past, both imaginatively and in terms of more conventional practices of seeking knowledge. This suggests that practices of both retrieving 'knowledge' and creatively imagining are difficult to disentangle in encounters with the past; both are part of a spectrum of engagements underpinned by an emotional desire to relate to the past of the home. But people also showed awareness of the dangers of allowing themselves to dwell within a more unbounded sense of speculation or embellishment about past residents, and of the reasons why this was appealing to them.

References

Berger, J (2007): 'Which Prosthetic? Mass Media, Narrative, Empathy, and Progressive Politics'. *Rethinking History* 11.4: 597–612.

Blackman L, Cromby J, Hook D, Papadopoulos D and Walkerdine V (2008): 'Editorial: Creating Subjectivities'. *Subjectivities* 22.1: 1–27.

Cameron E (2012): 'New Geographies of Story and Storytelling'. *Progress in Human Geography* 36.5: 573–592.

Cannell F (2011): 'English Ancestors: The Moral Possibilities of Popular Genealogy'. *Journal of the Royal Anthropological Institute* 17: 462–480.

Harvey K (2015): 'Envisioning the Past: Art, Historiography and Public History'. *Cultural and Social History* 12.4: 527–554.

Robertson I (ed.) (2016 [2012]): *Heritage from Below*. New York: Routledge.

Schofield J (ed.) (2014): *Who Needs Experts? Counter-Mapping Cultural Heritage*. Abingdon: Routledge.

Shuman A (2005): *Other People's Stories: Entitlement Claims and the Critique of Empathy*. Urbana: University of Illinois Press.

Webber S and Mullen P (2011): 'Breakthrough into Comparison: "Moving" Stories, Local History, and the Narrative Turn'. *Journal of Folklore Research* 48.3: 213–247.

Weisman B (2011): 'Front Yard, Back Yard: Lessons in Neighborhood Archaeology in an Urban Environment'. *Present Pasts* 3: 19–25.

Yarrow T (2019): 'How Conservation Matters: Ethnographic Explorations of Historic Building Renovation'. *Journal of Material Culture* 24.1: 3–21.

3   Presences of the past
Energies, auras, ghosts

The relationship between historical knowledge and imagination is made more complicated by more-than rational experiences and beliefs involving the continuing 'presences' of the past. In this chapter I will examine their impact upon people's relationships with their homes.

Anxiety about 'knowing' the past

Some participants, as described, showed a degree of reluctance to know previous residents as individuals – in case, for example, they were disappointed by them. Karen in her Victorian terrace in north London expressed frustration that her house had been modernised by a building company before she arrived, making the presence of the past less tangible because so few older features remained. But she was also relieved. She didn't want any extra aids for visualising past residents. By imagining time as non-linear, she suggested how 'crowded' the place would feel; this might become somewhat uncomfortable:

> People talk about four dimensional space as in three dimensional space and *time*. *Three* dimensionally, they did it exactly in the same – you know, they *physically*, you know, had a long day with the children, sat down here in front of the fire, did their sewing. And I did that – *do* that too. Feels like almost – you know, if you took out *time* – which obviously you *can't* – but if you took *time* out of it, we'd all be sat on each other's laps. You know [laughs]. Which I guess is a concept of ghosts, taking out time, if you believe in ghosts. If you see a ghost, not that I have done, they are *invading* – from another time.

Such a fantasy of 'taking time out' suggested a rather claustrophobic sense of co-habitation with the past, full of strangers taking up space in her *own* space at home. Her feelings of unease were dramatised as she focused upon the private, messy, events which might have taken place within these same spaces:

> It's almost a bit *gross* … [Looking at census records] presumably some of these children were born in our bedroom? [Laughs] And things like that.

Presumably anyone who was born whilst they lived here was born in our bedroom. Because they wouldn't have gone to hospital. Which is slightly gross, but quite fascinating ... The whole kind of [pause] very *physical* and *personal* thing that's happening in *our* – space. That's kind of – of *nice*. But a bit weird. Don't know. It feels like *ghosts*.

Imagining previous inhabitants in the private spaces of the home can, if the distance of time is removed, cause a visceral reaction; the past becomes too close for comfort.

Pam in the Yorkshire farmhouse also reflected on why she did not wish to discover more detailed knowledge about past residents or imagine their stories. They needed to remain 'faceless', distant generic people, in order not to give them too much imaginative *power*: 'If you start inventing or identifying their stories and giving them personalities, then your house becomes haunted by them. Then they start *occupying* the space *with* you'. She preferred to believe that they lived in parallel universes in the house – and that these should not come into contact:

I live in my conception of this house and they live in their conception of this house. And there are crossing points. But they have a separate continuity and existence from my continuity and existence almost. And I don't want to bring them together and make them overlap.

The threshold between the 'world' as lived by previous residents, and her *own* time needed to be kept separate rather than shared. She was very clear that she will only imagine past residents from the outside, from her own perspective – a refusal to inspect the 'other' in ways that might acknowledge them, thus potentially inviting them in: 'I don't have any sense of who these people were as individuals. I'm not trying to understand them from inside them, no. I retain a resolute – my *own* perspective on them'.

This need to control and place boundaries upon her imagination was motivated differently to those who were concerned that their imagination would make them stray beyond the 'knowable'. Sara in her west London maisonette shared this anxiety that her imagination might not just create its own false reality but stray into *other* realities. Like Karen, her flat had also been gutted by builders, but she fiercely guarded her space from any potential 'co-habitees' from the past:

This is *my* space. The space has changed. Just as it was their space for a period of time, now it's *my* space. So I don't really want to be – kind of *inviting* in too much speculation about, you know, who was here. Or starting to *imagine* them. Really. It just doesn't feel – it would feel like an *unwanted*, um, kind of thing, really.

Again, a clear temporal boundary between *then* and *now* needed to be maintained to ensure the past did not invade. Admitting to a fear of ghosts ('I don't read horror stories. I don't go and see those kinds of films'), she needed past residents to remain 'imaginary', as only ever existing in her *mind*. Her anxiety was reinforced by the fact that all she could find out about their lives was 'in this space now' – within her own home: 'That was *then*. And it's passed. And I don't want to get into imagining these *imaginary* people, because that's all I know about them – in *this* space … I don't want to go too far down that road actually'. Imagining the ghosts might lead to offering them 'hospitality'. At this point the conversation shifted to a focus on the more recent history. Thinking about past residents who might still be *alive* felt more 'comfortable', she admitted: 'I'm just thinking that maybe kind of – looking at the 60s and 70s might be a more – sort of *comfortable* place [laughs]. I know it's kind of irrational in a way'.

Visualising, seeing, feeling past residents

A further, but distinct, way in which participants question the overlap between knowing and imagining concerned a slippage between what is 'seen' in the mind's eye and something with more of a distinct 'reality'. Lydia, living alone in her ex-co-operative flat in London, remarked as she glanced out of her living room window: 'I'm seeing people in clothes of that time, in these streets, which I'm seeing as much darker. And yet, they wouldn't have had all these trees then … I'm going to their lives and then also their deaths. You know, where we're speaking about that person there [in census records]'.

The statement seemed clear and dogmatic – she was *'seeing'* these people. She offered no further explanation, but appeared to reinforce the vision by expressing her surprise at how dark it felt, given there would have been fewer trees.

Other participants offered more vivid descriptive details of their visualisation of past residents, appearing at times to slip away from an exercise of the imagination to some different terrain of encounter. Susan in south London described her fascination for the woman who built her house by following a familiar fragmented trail of knowledge accumulated through archives and hearsay. She built upon this to speculate on the circumstances surrounding a dramatic event: the woman had been forced to give up the house by the land-owners because she had not followed strict building regulations. Susan then shifted her narrative to the woman's appearance:

SUSAN: She's going to be tall and thin and quite domineering I think. Because she would have had to have been, to have stood up to the [free-holder] Estate. I mean … to stand up to the Estate is a big deal because, even now … you know, they're pretty, obstreperous [laughs]. And I just think – oh my *goodness*!

CL: Why tall and thin?

SUSAN: I don't know … I have this sort of Edwardian image – I don't know why I think that … She's got some *presence*.

CL: To frighten the middle aged men? [Laughter]
SUSAN: Yes, yes, yes! Hopefully taller than them [laughs] ... I do see her like that. But she must have had such a strong character to have done that at that time.

Susan's image of the woman seemed to amalgamate an idea of her character and her era. Unlike the 'one-dimensional' photographs of the Victorian lady who took to her bed, her appearance matched in vivid detail the assumed strength of her personality, gleaned from her willingness to 'stand up' to the men controlling the Estate, even if she had ultimately failed.

Other descriptions are more detailed still. Abigail, in her Georgian flat carved out of a large London mansion, described her sense of two women in the living room. The women, she believed, were from the house's earliest time. At first, she seemed self-conscious about the vision, and wished to pass it off as another nostalgic image influenced by television costume dramas:

> I know this room – this room in particular has a history of real people being in here, linking in with the Jane Austen fantasy of what they look like. Now I did have a colleague who was a historian and I can remember there was a *Pride and Prejudice* being serialised on television and everybody was watching it. And we'd come in the next day and say, 'Well?' And she'd say, 'Now, that wouldn't have happened, *that* wouldn't have happened'. And she absolutely did deconstruct – certain aspects of the décor and explain why they couldn't possibly be reading in the evening or sewing or – it was very fascinating ... Maybe that's just simply me having watched something on television, imagining two such people standing there.

She quickly dismissed this explanation: 'But that *doesn't matter* because in my mind I have a picture of – I say *women*. I only relate to women in this, being in certain clothes and actually existing *here*'. This phrase 'actually existing *here*' contrasts with the previous story of the inaccurate costume-drama portrayal, allowing her to emphasise her intellectual understanding and thereby, paradoxically, the credibility of her belief that the experience was *real*. Her image might appear to conform to popular cultural understandings of the past, and the 'picture' of the women to be in her mind, but it was the 'actual' women who 'existed here'; the picture is also *there*.

To explain this further, she referred to another cultural reference – Tom Stoppard's play *Arcadia*. Juxtaposing the early and contemporary inhabitants of a stately home, this weaved between them as scholars attempt to unravel the stories of the earlier inhabitants. The play represented one story which was inaccessible in the present: the casual solving of a so-called impossible mathematical theorem by a brilliant young woman.[1] The play is in part a comment on the approximations of what can be known of the past, parodying one historian's competitive but contrived discovery about the connection of the poet Byron to the house. The final scene suggests, as part of a dramatic device, the spatial intermingling of the present and past – the kind of over-lapping that Pam in Yorkshire wished to avoid. Abigail recalled her response to the play as a way of conveying her sense of the presence of the past in her own home:

> Have you seen Tom Stoppard's *Arcadia*? The very last scene is when the modern ones are wearing clothes [from the earlier era] they found in the attic. And they're dancing but the people from the 18th century are also on stage and they're ghosts, and they're still there, their essence is still there ... We know the history, we know the girl died on her 16th birthday in the fire ... And then it comes together. They're there. They're still there. And truly, I mean it overwhelmed me because I was living here at the time and I thought, 'I *know* this. Yes, I know about this'.

She then offered this description of what she saw from her vantage point in the living room:

> I do see a couple of women here, just standing here, just two women standing here. I'm not seeing them now. But – I know one of them is – I think one of them is wearing a yellow dress. And they just seem to be standing here talking ... And they just seem to be talking quietly. I don't see any men here ... They're very pleasant. They're having a nice conver-sation. I know one of them's got flat shoes. I can see the front of one and the back of another. And the one standing over there, and I think it's this – this yellowy dress, she's got little dainty flat shoes on ... [It is] a fond image of these two ladies just chatting and as I say, I can only see one of them, the other's in a dark sort of outfit.

As others, her interpretation oscillated tentatively between imagining and knowing, but her position was to allow the image to *be*, to have its own reality:

> I've never put myself into a state of trying to sort of think it through ... I don't need to. I don't think I need to. I mean I don't need to *know* ... But I do see two women standing here.

She was content to accept her experience as a different form of 'knowing' that required no further research or explanation. But, finally, she also didn't like the women being described as *ghosts*; this belittled her experience. She proposed instead the word 'spiritual' to more accurately reflect her feelings in witnessing particular past residents enjoying a moment in a particular place in her home.

Elsewhere, Yolanda's experience of the first woman of her 1930s suburban semi, was, as described, also expressed as an intensely personal connection.

Here the encounter was less a passive observation and more a repeated sharing of a domestic routine:

> In the winter, I come downstairs and make a cup of tea. Take it back upstairs for us to have our cup of tea in bed. And when I walk up the stairs, and it's dark, there's a very – I have this sense that, it would almost certainly be the woman of the house – she probably did the same in the mornings ... And I've had that sense, you know, for a long time. But it doesn't happen in the summer [laughs]. It's something about the dark.

Her interpretation of this feeling of connection first toyed with ideas about metaphorical ghosts and stored memories:

> It's ghosts. I think it's ghosts. I think it has to be metaphorical, but it's the ghosts of the people who used to be, work, live, sleep in those places. But I have the sense that houses – [sighs] it sounds very odd, doesn't it? That they kind of *store memories*.

She elaborated that her feelings seemed to relate specifically to her house:

> So there's something particularly about this house. I never experienced it in any of the other places I've lived in, but there's something about this house. I do think of the original owners and I really, really wish I could meet them. And I can't because they'll almost certainly be dead now.

Her account, of (on the surface of it) a commonplace domestic practice, is ambivalent, focusing less on the imagination and more on the sensual experience triggered in the act of doing the same – an embodied, affective form of knowing. Indeed, she described her 'sense' and 'feeling' about the woman, beyond conjecture about her life, with the feeling specific to the house. The fact that it only happened in the dark mornings of winter adds specificity to the story, suggesting a hushed, muted atmosphere and the closing in of the senses. She expressed the quality of the experience in a similar way to Abigail, as 'spiritual':

> It's almost *spiritual*, I think. It's as if I'm – it's almost as if I'm actually doing exactly what she did. I'm retracing her steps. I'm literally, aren't I, retracing her steps, *every day*. However many times a day the woman goes up and down the stairs with another bundle of washing or whatever. But it's not *that* I think about. It's – actually, you know, I find it very difficult to put into words [pause]. It's something about the beginning of the day. Starting another day. It is *me*, and *that woman* – on the stairs [laughs]. There is a connection between us [pause]. Is the word 'ethereal' right, maybe? Obviously, it's intangible. [Pause] It feels like there's something

more than [imagination]. There's something more, that I can't really describe very well. It is this sense of retracing her steps. And not just *literally* – because that's what I'm doing, literally, aren't I? I'm walking up and down the same staircase that she walked up and down. It's about *living* – it's about – how – *what* she's going to do during her day, and what's today going to bring. I have no sense of a [ghostly] presence … She's alive … She's alive to me. In the same space.

For both Abigail and Yolanda, the experience concerned specific domestic spaces, but the relationship is more complicated than being just about either ghosts or the imagination. The next chapter will elaborate further on different senses of connection with previous residents. It is perhaps telling that in these cases, the connection is between women.

Encounters with real ghosts

There are further, less hesitant and more clear-cut descriptions of past 'presences'. As described in Chapter 2, Gemma in the town on the Thames believed she was haunted by a laundress living and working in her cottage around the turn of the twentieth century. Gemma didn't feel the need to *see* the ghost but is 'quite happy with her being *there*, right beyond my shoulder as it were. She's just out of vision, on the edge of vision'. Like any good ghost story, she tells hers with a degree of relish:

I went to tune a piano for a dotty old lady in H.W. And I roared up in the Morris Minor … And she said, 'Oh, come in, *both* of you'. And I looked around [laughs] and thought, 'it's just *me*'. And she said, 'We'll go and make the tea. You go and get started on the piano'. And there was nobody else as far as I could see [laughs] … And she said, 'Lizzie says you've got a temper on you'. I said, 'Well, yes. Sometimes'. And she said, 'You live in a funny little house along a long drive'. 'Sort of, yes'. She said, 'Well Lizzie used to work in your house' … And she described this woman: boots. Long grey skirt. Pinafore. Hair scraped back in a bun. Snaggle-tooth. So I thought nothing of it – 'she's a dotty old – *dear*. Never mind'. And – it might have been a year or so later – I came home and there was an old chap, *standing* at our gate, *peering in.* So I thought, 'nosey old' – I came up behind him. '*Excuse* me'. Made him jump. He said, 'I wasn't being nosy. My gran used to live here'. He was about 80 odd … And I said, 'Was her name Lizzie?' I don't know why I said it. And he said, 'No. Liza'. And I said, 'Did she have a snaggle-tooth and hair scraped back?' And he said, 'Yes'. And he went a bit funny, and I said, 'Do you want to come and sit down?' [Laughs] So he came in and had a sit down. *She* was the laundress. But she was a fairly feisty old – thing. She chucked her husband out as he was more or less a parasite. And he used to bring parcels of – obviously ladies' under things I would imagine – to be washed, when he was a boy. And he also used to have to bring her a jug of ale in. Quite regularly [laughter]. He used to bring a jug up from the pub. He said he used to have to walk up from The Hope [although] there were lots of closer pubs ... And I said, 'Things go missing', like jewellery. He said, 'Well, she liked nice things'.

Her initial scepticism about the ghost, followed by the apparent corroboration of the story with its satisfying details, allows the listener to simultaneously manage and suspend disbelief. The line between imagination and 'reality' is policed in such a way as to allow for a crossing over between the two whilst maintaining a degree of distance. Such an uncanny narrative, casually mixing the strange and familiar, is a common but overlooked form of performed oral history storytelling – part of the way anecdotes about experiences and beliefs related to particular local places circulate to offer rich, vivid layers of additional 'knowledge' to inherit or pass on (Bell 2012; Lipman 2014).

The ghost suggested a focus for research; Gemma wanted to know who she was. The census records offered a few possibilities of different 'Elizabeths', or shortened versions of the name – a point of discussion never fully resolved. But the ghost also acted to reinforce a sense of identification with the cottage. This presence from the past reinforced Gemma's own sense of presence in the cottage. By reflecting on their shared personal characteristics, she emphasised her own fit with the home:

> Sometimes when I sit down with a pint of beer, I think, 'Blimey, I'm turning into Eliza' [laughs] ... I get the *impression* that she was very *tidy* and everything had to be *just so*. And in my business I'm quite organised, but other than that, things just happen [laughs]. So occasionally I think she's *bristling* when the house is in uproar ... She's very *involved* ... Strangely, I think how we are quite alike. We're both quite feisty. Both like a pint [laughs]. And both run our own businesses – and do what we like, when we like. Which presumably *she* did.

Elsewhere, another description of a ghost dramatised a home's affective atmosphere. South African–born Jack lived alone in a small terraced house in Bristol, previously with a shop at the front. He had only moved in a few months earlier. It was still in the raw, neglected state in which it had been left by the previous owner. Jack described sensing a male presence which he presumed was 'sussing' him out as the new owner:

> It really sounds weird but I often think that there's a presence here ... somebody just sort of sussing me out and looking at me and thinking 'ah ha, this is the new – this is the owner, let's see what he's like'. It's not, 'oh this house is haunted and we don't want a human being in here'. Or whatever... I almost feel as though, that I'll either live with it or that it will go

away, you know … The house was built in 1835 so it's been here a long time so there must have been people who lived a long time here.

His assumption about the presence's anxiety about him seemed to reflect his own ambivalent relationship with his new home:

It's somebody who's coming here and I've got the sway over this haven't I? I mean I can knock windows in, I can knock walls down. I mean I can do untold changes or damage. So I think maybe it's just somebody who just wants to know what I'm going to do … I don't particularly want to make a connection with him.

The presence, he added, seemed to be 'backing off' as it 'realises … I'm not going to upset things at all' – Jack lacked the funds to modernise the house. But during our guided tour, he stopped and shuddered at a dark turn of the staircase near the top of the tall, narrow house. Here was a rather different experience of a morning perambulation than expressed earlier by Yolanda. He said:

I don't like this. When I come down in the morning I don't like this area. … The area just gives me the creeps when I come through there. … I hate it in the morning. I'm fine with the rest of the house but I come down here and I just hate it.

The uncanny atmosphere was reinforced by Jack's belief – knocking on the wall to demonstrate a hollow sound – that there was a 'hidden cavity' behind the wall, a further reminder that he was stepping into a space with its own history he had yet to be fully acquainted with. The hidden cavity is emblematic of the hidden places and secrets which connect a definition of the homely (*heimlich*) with a definition of the unhomely or uncanny (*unheimlich*); here the *knowledge* of the cavity reflects an *unknowing* of the home (Lipman 2019). The presence of a ghost also offered a different kind of access to the home's past, but here its identity, beyond a sense of its gender and intention to haunt, was vague and his desire for distance from it, unlike in Gemma's experience, was reflected in his *lack* of interest in searching the census records for more information.

In other examples, the census records themselves could trigger uncanny feelings. In the east London square, Carol had explained that she had little emotional response to the census names, describing her 'bricks and mortar' approach to the homes she had lived in as a form of 'serial monogamy'. But this attitude was dramatically challenged towards the end of the interview. Returning from a guided tour of the house, Martin and I found her sitting quiet and still on the sofa, staring down at a copy of the 1901 census records. She said quietly: 'You know. Something has just made my – my skin is – *creeping*'. She continued:

When our daughter was really small – in the kitchen, she used to say to me: 'Mummy, I can see two girls in Victorian dresses' ... And I used to think it was nonsense ... I just thought it was those make-believe friends that children have. They had bonnets as well. She said they had bonnets ... I did think, 'Oh how odd' ... And I didn't want to pursue too much of the conversation because I didn't want to make her feel – *Now* I'm thinking: 'I wonder if that was [looking at census] Eugenie Williams or Miriam Williams floating around the house?' [Laughter] ... *Now* I've got names [laughs] ... It made my skin go funny ... Fascinating. I won't sleep tonight if I start thinking about it! It'll be floating around my head [laughter].

Carol had the classic visceral response – creeping skin – as she described the moment the census data triggered a memory of, and appeared to *verify*, her daughter's earlier ghost sightings she had dismissed as imagination. The story emerged out of and reinforced her own familial history of being *in* the home. Indeed, the different ways in which the past became 'present' (or as in these cases, a 'presence') both dramatised and at times influenced people's emotional responses to their homes. The ghosts also demonstrated the way people explored and explained the relationship between what is knowable or unknowable about the past, in terms which extend definitions of what is, indeed, considered 'knowledge'. And the ghostly encounters also offered more vivid, visceral and tangible ways for people to articulate their unexplainable feelings about particular spaces of home – giving these 'others' a clearer shape, if not always a 'face and form' (Ahmed 2000:3). Indeed, the immediacy of specific presences of past others – some of which are gendered or granted their own human-like responses such as anxiety or annoyance – offers space to reframe how relationships with present selves might be shaped.

Unhappy homes

The ghosts also bridged a gap between a sense of the home's past and the way the home now 'feels', its atmosphere or aura. Indeed, many participants assumed that the older a home, the more likely it would be affected by the events and people who had lived in it before. Emily and Derek, for example, pointed to the 'lovely feel' of their 1920s arts and crafts house deep in London's Metroland. This home didn't share the same 'time depth' of Emily's old, charismatic childhood home, implying that this accounted for its friendlier feel. In contrast, the old family home had a 'reputation of being one of the most haunted houses in the country':

There were drafts in funny places. It was an odd house. It really was. Certain parts of it could feel uncomfortable. ... The late 17th century painting, 'Duchess of Richmond' *terrified* my brothers. And it was even worse for my mother – she placed her face to the wall because they thought she would jump out of the picture.

But her childhood house, she added, laughing, also had a '*contemporary troubled atmosphere*' as there were 'a number of *troubled people* living in that big house, together'. Derek added: 'And I've read the diaries [laughs]. If people really don't want future generations to know about them, they really shouldn't write diaries'. And yet Emily wondered how far energies of places might *linger*: 'There are houses that just plain *feel bad*. And whether you believe in ghosts or not – it's an interesting one. There are *feelings*'.

This ambiguity as to how far an atmosphere of a home is based on past events or current ones reflected another experience of home. On the fringes of east London and Essex, Sandra rented a flat carved out of a large old manor house. This building has an iconic status in the neighbourhood, partly because of its imposing presence, standing in lonely defiance in a neighbourhood of modern suburban houses, but also because of its strange appearance, with a castellated section of roof, a bell tower without a bell, and a fierce-looking gargoyle glaring from a corner wall: 'It is just the weirdest place, a mysterious one off … with no connection to everywhere else'.

Talking to the elders in the local church, Sandra discovered that the building had long been considered to be haunted:

> One man said that when his children were *children* – they're now grown up – they said that this place was haunted. They obviously ran around and played everywhere, but they would never go in there. You know [laughs]. There are things you hear, there are little snippets I hear.

But the strangeness of its appearance and its local reputation are reinforced by how the building had been treated more recently. The house had fallen victim to local infrastructural change, finding itself adjacent to a busy motorway, its incongruous, liminal placement in the landscape reinforcing its isolation and contributing to its economic demise. Sandra described the neglectful absent landlady who did not maintain the property and treated its high turnover of tenants with little respect. Sandra, at ten years in the house by far the longest resident, recalled her friendship with ex-neighbours in an adjacent flat:

> About three lots of neighbours ago, they were both into what you might vaguely call spiritual things, or – not witchcraft but psychic things. Not in a bad way. She used to say, 'I do spells, but I never give *bad* ones' [laughs]. But they were absolutely convinced that this was a *cursed* place. And that is why so many bad things have happened … They said there's a jinx on it, that bad things would always happen. They said they could feel bad *vibes*. I said, 'Well', I thought that *all* the problems, a hundred per cent of them, were caused by the *landlady*. Practical things that don't need to be like that. If someone doesn't do a repair, that doesn't mean the place is cursed, you know.

Her response to vague claims of a curse shifted the blame for the 'bad vibes' to the realities of the home's neglect and those, like herself, trapped in such places due to wider socio-economic circumstances. Despite this, the neighbours had put doubts in her mind:

> But it didn't make me feel that good. Particularly because they said that they had cursed the *flat* – before they went ... as a sort of 'goodbye. We don't like you. We're leaving'. A sort of two fingers up. ... They made sure they isolated it so it didn't come towards *me*. And they had a lot of trouble getting people in there. Either people didn't stay, or didn't pay the rent, or it was just void. ... I couldn't deny the fact that *so many things* have happened here. Year on year on ... I mean, I did once have this completely silly idea. ... You know how in some cultures people would get someone from the church to come and bless their house? I even did think, 'Well, why don't I do that?' You know, I mean ... And then I thought, 'Oh come on, you're getting ridiculous now, you know. If they had never said it, you would never have thought it'.

The questioning of her home's bad atmosphere and neglect, focused on the rumours of hauntings and curses, reflected Sandra's own vulnerability in the flat and her anxieties for the future. The house's pasts, the accumulated bad events, only highlighted its present neglect and the social power geometries which impacted the life of residents.

Sandra's story also had echoes of the experience of Lydia in her ex-co-op flat. Lydia described how she was suffering from 'anticipatory anxiety' due to major maintenance problems which the freeholder, the local authority, had been slow to repair. She had wondered if the 'bad luck' she was experiencing in the flat was caused by something that had happened previously:

> I'm interested in what the house might hold in terms of what's gone on here ... As soon as an idea comes into your mind that there's an atmosphere here – there's almost like an *opening up* to history ... A different level of perception.

But she wasn't sure if the flat was haunted or if it was something 'in her head', explaining that the 'unsettledness' she experienced in the place might have been 'in *me*, in regards to the home, as well as *the home*. And it's always very difficult to sort out what's what ... where one thing would end and another thing would begin'.

Energy accumulates or winds down

Lydia did believe that people leave 'energy auras' or 'trails'. She imagined the ways these might linger:

I'd love to see those traces … what was left of your energy and all the energy of the people who have passed through these rooms, as a kind of picture, of how it might be. That must be what we're *in*. That must be partly what atmosphere is composed of … It's all transient. It will go back into the ether. … Ether is the only word you can use, and ether isn't even a real *thing*.

She wondered how long such traces or trails might last; previous 'bad energy' in her flat following a relationship break-up appeared, she said, to have 'neutralised' over time.

For some participants, a home's 'aura' and 'atmosphere' is tied very directly to its materiality. Sandra and Lydia suggested this in relation to their home's neglected physical state, and Rita, as described, reflected on her family home's aura in relation to its smell, old furniture, curves and slants. In turn, social events are also considered important – the fact that a home is only defined in relation to the people living in it. Carol in the east London square told an anecdote of a friend who went abroad for six months, leaving her house empty. When she returned

she said the house felt *dead*. It didn't feel like a home anymore. There was no *atmosphere*. And she said to me – she'd been back for about a week – she said: 'I feel that the house is sort of coming *alive* again' … The energy that people make within a property.

Carol related this idea to Martin's mother's house, whose partner lived next door. They had taken the fence down between the gardens and he always came round at nine o'clock every morning:

When she died he bought the house … and it's no way, shape or form a *mausoleum*, but the house is pretty much – is *exactly* the same as it was … He's not changed it. He's not changed a thing. But that's his way of dealing with – I think he thinks she's gone on holiday – that's being too simplistic because obviously he knows she hasn't … Although her clothes have been removed, the house – it's still got an energy in it although it hasn't got her presence. I don't feel her there.

Although her point was that it is *living* people that make the 'atmosphere' of home – the woman's 'presence' was no longer there, but the house still had an 'energy in it' which she couldn't explain.

Adam in east London also wondered whether the 'spirit' of a place is absorbed through its 'fabric' or through the 'people':

I can imagine living in a Romany caravan and absorbing some of the spirit, the people went there before … Like is it just the fabric or is it – the people and the history? It's the people as well … I don't believe there's any ghosts here. But I believe that you can leave an imprint on the place, for good or evil … If something really nasty went on here, I'd get it exorcised or something, you know. If I really felt bad about it.

For most participants, the home was considered a container of lingering past effects, complicated by the fact that it continued to be a place of *present* and ongoing occupants and events. In many cases, contemporary circumstances appeared to contribute to its accumulated affects – Yolanda's 'stored memories', which, for Susan in south London are 'layers of experience' creating a 'patina of memories'. Indeed, Lydia's belief in transient 'traces' constituting atmosphere or ether became a core consideration, and many participants wanted to know how long the *energy* from a person or event lingered. Most also assumed, like her, that it 'wound down' over time. William, reflecting on past homes he had lived in, described this in terms of the distance of time. In older homes, the people were now 'too far away' to have any 'connection' with the present. This belief led many participants to reflect on what events might have occurred in their homes. Are some events more prone to creating negative or positive atmospheres? Do more extreme events leave a particular mark?

Perhaps inevitably, speculation turned to death; there was an understanding that, like births, these would often take place at home. On the one hand, for some, this was not in itself a cause for concern because it was considered a 'natural' phenomenon, a part of life. Adam illustrated this, batting away what might have become an unsettling story: 'An old guy came up … and said to me, "the old boy upstairs, you know, the day he died in the bedroom upstairs, the tree fell over". But I'm okay with people dying. You know what I mean?' Others share the same sentiment. In her east London Victorian home, for example, Cathy described how being born, dying and having 'ups and downs' in a home was 'the natural course of things so it's fine. In a house *everything* happens'. But this was contrasted with death in 'unfortunate circumstances' which might, she admitted, 'spook' her, leave 'bad energy' – another reason to avoid too much detailed research into the home's past:

I'm not sure how I'd feel if there had been a *murder* here. I mean, if there was somebody who died in very unfortunate circumstances – that might spook me. You know, life comes and life goes in any home, but – bad energy from something *that* traumatic that happened, you know … The anguish lingers. If I knew about it, that's when it would worry me. So not to look too deep [laughs].

She described a 'place in a street' where she had

witnessed something really awful happen, and it still gives me a shiver going down there. It could just be the memory has captured you in it, these things grab hold of you … But it wouldn't change my relationship with the *house*, maybe the air or the room [laughs]. Because the house can't be held responsible for what happens in it ... No, it's more *ether* I think, for me.

Cathy separated the house itself from the 'lingering' effects of events within it. It is not intrinsically the house's fault – here an innocent witness, or at least a neutral vessel, within which the dramas of life play out. And it is only an ethereal form of lingering – something insubstantial. This personifying of home as having a *tangible* physical presence which can be dissociated from the intangible reverberation of events within it helped her to emphasise her primary, benign relationship – with the home itself.

Julia, in Keats' home, was equally philosophical about death in the house, but questioned how she would feel if she discovered a past murder:

> Death is something that comes to us all, right? If I'd known a murder had taken place, you know, that somebody had come in here and murdered someone, I think I'd feel bad about that [because] of the violence and distress involved.

She also described a past 'terrifying' experience of her own, and how this gave her a personal insight into the emotional intensity involved in acts of murder:

> I've experienced something similar myself ... Somebody did try and kill me once, and I did survive right ... But so I know how utterly terrifying that is and so I suspect – if I knew that something like that had happened here I would feel – I feel I would have to make a conscious effort to overcome it, you know, as an element of the environment ... Even though I'm not superstitious, you'd still be wondering, 'Is there some kind of weird vibe left over?'

This fixation on murder became a repeated theme in the interviews. Samantha in the north London maisonette differentiated between different kinds of murder, with 'somebody being kept here for a long time in unpleasant circumstances' being 'harder to deal with' than a 'sudden' murder – the implication being that negative energy would have longer to accumulate in the former case. She reflected: 'How long does it last for? I didn't realise I was that superstitious. The kind of feeling of emotion in a place'.

Nigel, living with his wife Jane in a west London suburb, was also concerned about the effect of time on events. He said: 'You can get some quite *dramatic* history of the house, couldn't you? Murder – it wouldn't be a very nice thing to find out. There's always the risk'. Jane added: 'I'm sure the longer it goes *back* the more there's *something*. It's like looking back at your *own* past, isn't it really? Normally you'll find something that might be a bit – slightly upsetting'. Nigel speculated that, if this is the case, the *timing* of the murder mattered most – but so too did the *place* where it occurred:

If it was something like in 1920 I wouldn't be worried, but if it was – you know, you bought the house five years ago, or last year, and someone was brutally murdered. If there's been a murder in the bath and the bath was still *there* and you found out that someone was murdered in the bath, you wouldn't be very happy about that, would you? If you're having a bath and the bath was used for a murder – you'd go out and swap it.

The reassuring idea that a murder would no longer have an impact on the home's atmosphere after a period of time is supplanted by his response to a different kind of anxiety around inherited *objects* – an old bath still in use might harbour a dark secret. Removing items of furniture might be easier than changing atmospheres, but it is noticeable that particular attention is drawn to discomfort around *certain* objects – in these accounts, usually baths and beds. These are used in the most bodily intimate, and therefore exposed and vulnerable, ways. In east London, Joy recalled asking the landlord to remove the beds from her semi-furnished rented flat:

I didn't fancy sleeping in someone else's bed. … To actually *sleep* in someone *else's* bed, I mean [shakes her head] … I regard sort of my bed as my personal – it's *me*. It's *mine*. It's a *private* space. I just thought, 'go away' [laughs].

For many participants, however, reassurance was to be found in the belief that haunted homes are always *unhappy* homes. Those who believed that theirs had a good atmosphere were less concerned about any psychic residues from the past. Adam in east London explained that his had a 'really nice feeling … just like nobody had died a violent death here. Nobody had been unhappy here, you know'. Martin repeated the claim: 'I only feel that it's been a happy house. I don't feel anything terrible ever happened here'. Abigail in her east London Victorian terrace also described the 'positive vibes' in her flat: 'here I have picked up nothing but peace'. And despite the presence of a ghost, Gemma is also clear about its 'aura': 'I think it has an – *aura* – I don't know. Everyone who comes in says it's very welcoming. That it sort of *envelopes you.* Somehow … *Atmosphere*, I suppose I mean'.

As described, concerns that they might discover something about their homes became a clear reason for some people *not* to investigate their home's history too deeply. Nigel, for example, shared Cathy's anxiety about finding something in his home' past, saying: 'It is interesting to some extent, who lived here before, but – I think I don't want to think about it too much'.

For William, this issue had become more immediate, having discovered through genealogical research that murders had taken place in a previous family home in Croydon, as he dramatically related:

Aged 25 in 1908. In a lower middle-class semi. Literally went like this [claps his hands once] and *murdered*. I did all the research. … I want to walk in to that upstairs bedroom where the two small beds were, where the two women – Roman Catholic – after lunch. ... They would lay like this, with the rosary. And, so the newspaper clipping said, he want mad, and he went in. Shot them very quickly at the side of the head. His sister was 30, unmarried. His mother was 58. Just turned around and shot the two family dogs, went to the dressing room and left a note, 'Please, please bury us with *him*'.

Out of curiosity, he had taken a trip to the street, armed with this gruesome knowledge he presumed the current inhabitants were unaware of. This set up a dilemma: what to do – or what *not* to do – with such knowledge? He expressed his unease in completing the tale:

> I felt shy in the street because I felt, 'this is bad manners. You're taking this photo without knocking on the door and saying, "May I take the photo?"' I would love to go and knock on the door and say 'Oh by the way'. But I thought, 'There's no way I can do that'. Oh no. It may well *upset* those people. I mean – *no*. I couldn't do that.

Figure 3.1 A gargoyle-like green man figure on the corner wall of Sandra's home.

Conclusion

This chapter has extended the exploration of people's experience of historical knowledge and imagination within their homes by paying attention to superstitious or more-than rational beliefs, arguing that these constitute an important and under-explored element of many people's engagements with the 'presence' of the past, be they imagined, felt or experienced (Bell 1997; Bell 2012; Holloway 2010; Lipman 2014; Peeren 2010; Royle 2003; Vidler 1992).[2] For some, beliefs which suggest different forms of 'presencing' of the past influence what people wish to know or not know of their homes' histories, again at times requiring strategies of distance – of reinforcing the gap between *then* and *now* – out of concern that the past as a non-linear construct might press too intimately upon the temporalities of the present. People's sense of presence sometimes created feelings of claustrophobic unease, but the imagination itself was also viewed with anxiety, considered at times powerful enough to 'conjure' up rather than just re-create an idea of the past. People also grappled with the line between appearing to *see* something – in the sense of a detached observation of things separate to self, imaginatively *visualising* the past as a form of costume-drama nostalgia conforming to cultural normatives, and somehow *entering into* the presence of the past or to be *enveloped* by it. In some cases experiences were granted greater personal meaning or impact – as something 'spiritual', a sense of connectedness with specific past 'others' who lived in the home. In these cases, it is noteworthy that both past others and present selves are often women, such encounters replicating not only traditional gendered associations of domestic routines and 'housekeeping', but adding insights into the ways ghostly others themselves are often stereotypically gendered, with female ghosts assumed to be unthreatening, vulnerable or caring (Lipman 2014).

For a few, ghosts were allowed to be 'real' in a taken-for-granted way, and were seen to deliberately interact with residents, making their presence 'known'. Beyond the figure of the ghost – as a narrative device, a figment of imagination, an aspect of reality, or a form that crossed between all three – there was a marked reiteration of a more general belief in broader 'presences' of the past. These were variously described as being, or creating, a home's atmospheres, lingering energies, auras or ethers, assumed to be an affective accumulation of and from the evolving events taking place within the home. The present is enlivened by such forms of affect, which become mapped onto the home as if objects in their own right with evolving biographies that can be traced, winding up and down in response to degrees of reverberation between *then* and *now*. These atmospheres and energies might at times have reflected participants' own emotional responses to their homes, but they also offered a particular form of access to the past which, on the one hand, offered more possibilities for expressing relationships to the home in broad terms, whilst on the other suggested a need to deal with the challenge of sharing home with a range of past and present selves and others who are granted degrees of agency to act.

Such beliefs and experiences make the project of relating to the home's pasts rather more complicated than that described in house history literatures. It forcefully suggests how more-than rational encounters continue to be ubiquitous within contemporary, Western, and so-called 'ordinary' or 'mundane' everyday spaces. In the light of this, those already critically evaluating the usefulness and unintended consequences of those official pan-global schemes to distinguish between 'tangible' and 'intangible' heritage objects and practices should avoid too narrowly defining what counts as 'heritage'. Specifically, we should reflect on the relationship between these 'intangible' experiences and feelings and the more tangible things, spaces and locations out of which such encounters emerge. Forms of intangible cultural heritage which embody more-than rational 'realities' and responses are, of course, difficult to capture, explain and track – challenging for those who believe we need to be more, not less, prescriptive in understanding how the past is or should be valued (Jones 2010; Kirshenblatt-Gimblett 2004; Watson 2018; Rose 2011). We should avoid imposing our own cultural assumptions or reducing such experiences and beliefs to only so-called 'rational' explanations, whilst attending to the situated contexts in which they play out. By allowing people the opportunity to share the full range of their responses to, encounters with, and beliefs about, the past, we can gain further insights into how the home's pasts are valued, reflected upon and felt, and how they impact on feelings of belonging to home. This offers a foundation for exploring, in the following two sections, more specific responses to the social and material histories of people's homes.

Notes

1 The theorem was first established by Pierre de Fermat in 1637, who, writing in the margin of *Arithmetica*, famously claimed he had a proof but the margin was too small to write it. The proof was finally published by Andrew Wiles in 1995. I saw the play on the same day that the announcement was made that it had been solved, giving the experience an ironic frisson.
2 I use the term 'superstition' without implying a derogatory meaning, a point of debate for Western folklorists (Goldstein *et al.* 2007; McNeil 2013).

References

Ahmed, S (2000): *Strange Encounters: Embodied Others in Post-Coloniality*. London: Routledge.

Bell, K (2012): *The Magical Imagination: Magic and Modernity in Urban England 1780–1914*. Cambridge: Cambridge University Press.

Bell, M (1997): 'The Ghost of Place'. *Theory and Society* 26.6: 813–836.

Goldstein, D, Grider, S and Thomas, J (2007): *Haunting Experiences: Ghosts in Contemporary Folklore*. Logan: Utah State University Press.

Holloway, J (2010): 'Legend-Tripping in Spooky Places: Ghost Tourism and the Infrastructures of Enchantment'. *Environment and Planning D: Society and Space* 28.4: 618–637.

Jones, S (2010): 'Negotiating Authentic Objects and Authentic Selves: Beyond the Deconstruction of Authenticity'. *Journal of Material Culture* 15. 2: 181–203.

Kirshenblatt-Gimblett, B (2004): 'Intangible Heritage as Metacultural Production'. *Museum International* 56.1/2: 52–65.

Lipman, C (2016 [2014]): *Co-Habiting with Ghosts: Knowledge, Experience, Belief and the Domestic Uncanny*. London: Routledge.

Lipman, C (2019): 'Living with the Past at Home: The Afterlife of Inherited Domestic Objects'. *Journal of Material Culture* 24.1: 83–100.

McNeil, L (2013): *Folklore Rules*. Logan: Utah State University Press.

Peeren, E (2010): 'Everyday Ghosts and Ghostly Everyday in Amos Tutuola, Ben Okri, and Achille Mbembe'. In M Blanco and E Peeren (eds.) *Popular Ghosts: The Haunted Spaces of Everyday Culture*. New York: Continuum: 106–117.

Rose, M (2011): 'Secular Materialism: A Critique of Earthly Theory'. *Journal of Material Culture* 16.2: 107–129.

Royle, N (2003): *The Uncanny*. Manchester: Manchester University Press.

Vidler, A (1992): *The Architectural Uncanny: Essays in the Modern Unhomely*. Cambridge, MA: MIT Press.

Watson, S (2018): 'Emotional Engagement in Heritage Sites and Museums: Ghosts of the Past and Imagination in the Present'. In S Watson, A Barnes and K Bunning (eds.) *A Museum Studies Approach to Heritage*. Abingdon: Routledge.

Part II

Past residents at home

4   Connecting with the past
Domestic genealogies

The act of sensing continuing presences of the past, or of believing such presences might continue, generated a range of responses, from feelings of spiritual connecting to feelings of anxiety. The latter, in turn, could lead to defensive strategies for maintaining distance, limiting how far the past is allowed to enter consciousness through the imagination or the pursuit of knowledge. Acknowledging the presence of 'others' challenges continuing ideal tropes that insist on the home as a private and familiar set of spaces and practices (Blunt and Dowling 2006). Indeed, the intensity of some embodied and emotional responses to past others suggests a distinctiveness about encountering the past within the *home* in comparison with encounters at other historical sites, in terms of degrees of power to influence senses of belonging and feeling 'at home'. Over the next chapters, I will explore further the different ways in which knowing or imagining the home's pasts might enhance or challenge senses of belonging through encounters with others who are more – or *less* – *similar* to ourselves. How are past others excluded or incorporated as part of the ongoing homemaking project (Varley 2008)? How do ideas, beliefs and values related to the perceived differences and similarities between past others and present selves inform this project? In this chapter, I will focus on people's responses to earlier past inhabitants, followed in Chapter 5 with an exploration of engagements with more recent and future others. The chapter begins by reflecting on the implications of a broader sense of being *part of* the history of a home, before moving in turn through examples of how people seek to make connections, or manage differences, between *now* and *then*.

Extended perspectives

Firstly, many participants made an assumption that an understanding of their home's past would enhance their emotional attachment to it. Gemma made this point: 'The more you know about your house, the deeper your connection grows', and for Cathy, it 'extends your sense of belonging'. This became important when senses of connection were lacking, such as for Susan in south London, who still pined for her last home which she loved but left for practical reasons. Having got to know the history of her house, she liked it 'more than I did'.

But for others, such as Abigail, interest in a home's history is an *outgrowth* of an existing emotional connection – in her case, a love of its original Georgian architecture. She readily placed herself in its history: 'I'm just a continuation of this census list', she said, looking at the records. Joy, in her rented flat in an area of east London she has lived all her life, said her interest related to her broader memories and knowledge of local history which reinforced her sense of connection to her home:

> I think it all ties in, not particularly because I'm interested in the history of the flat or of [the borough]. I'm interested in who and what – you know, what happened to the place where I'm living, who lived here and – where it all *connects through*.

For many participants, looking at the records reminded them of their relatively *temporary* lives in their home. This became a trigger for a sense of sharing home with past and future others across time which, in turn, offered a context for feelings of humility. Dan in east London described his sense of merely 'borrowing' the use of his house, and how this challenged a feeling of living in isolation from others:

> Rather than just living in your own snapshot world, it's nice to put that in the context of other people who have been within these walls. ... You could argue that I'm just here temporarily, that this house will outlive me. I'm just sort of *borrowing* it for a period of time. Legally I'm not, but affectively I am.

Julia described how knowing the history of her home offered a 'sense of connection' with the past which made her 'feel more grounded': 'You sort of feel a sense of '"Right", this is what I'm doing, but this is what happened before'. But she was anxious that this desire to place herself in the broader 'sequence' of lives might be seen as narcissistic:

> It's interesting to think, 'Well, I live here, so who else lived here coincidentally?' Or whatever, you know. Not so much that I *own* this place maybe, but more, I *happen* to live here. I live here so it's really interesting to know where I fit into the sequence of people who lived here, you know – where you kind of fit in, historically.

Ben in south London repeated the belief that people leave their 'imprint', but by this he meant more a contribution to the accumulating identity of the house *itself*, contributing to something more than the sum of its individual parts. The imprints are not isolated from each other; the house is characterised as something organic, a fluid depository:

The idea that it is a sort of little *mini organism*, I suppose … If you think about the people who have been here – or who will be here – then you know they have all left their little *imprint* … This is a depository actually, sort of a place where all these people like a swimming pool have come in and out.

The home perceived as a container of transient residence accentuates its temporal sweep. The lives lived under its roof are given positive value, often configured as part of a *'continuity'* – a word repeated as shorthand for a generalised connecting together which strips away differences, always granted positive value as associated with feelings of emotional grounding. It is within this 'pool' of continuity that things repeat, sustain, and (even if only imaginatively or in retrospect) are *shared*. Chris in the West of England described the emotional quality of such an idea: 'It's a reassurance, and sort of comforting to think that you're just part of a bigger picture'. Leah in south London expressed this dramatically:

It's a sense of your place in the world I suppose … It's just knowing that you're not the beginning or the end. You're just a part of it, and it will continue. And things around you – this house will be here after I'm dead, and after Alex's dead. And there's been loads of other people that – you know, the house is like *bigger* than one human being … It's just an incredible kind of thing to think about. It's one of those things that blows your mind, like when you think, 'oh … is the universe bigger, where does it go on?' When you live there and think about it, you think, 'wow, that's *mad*'. You can't – it's incomprehensible, isn't it? You think – *you* and your characteristics and where you *are* – you are not unique really because it's like there's a big *line* that goes back and it will go forward as well.

The feeling of awe about connecting to the 'big line' leads to a related sense of 'permanence' – quite a claim considering most places in which we live are not *that* old and are flimsy, human-made structures that can come down quicker than they went up. Here, identifying with something *greater* than oneself enhances the significance of a sense of being at home. Leah later described this in relation to her own *insignificance* (as a 'little dot' in time): 'It's two ends of the scale, because I still am *there*. I still am very connected. But I'm just *that*, a little dot. Very connected and very belonging and everything else – but a little dot'.

Beyond these more abstract feelings, there are also more specific triggers for feelings of connection to the home's material and social pasts.

Haptic continuities: original fixtures

Responses to *original fixtures* offered the most uncomplicated anchors for enhancing senses of continuity and connection. For Leah, it was the old trees in the street. She had found a photo taken in 1901 showing a horse and cart and two large oak trees – the last survivors of ancient woodland cleared to make way for urbanisation – which were 'just the same' *then* as *now*: 'It's just amazing that a hundred years ago the same two trees were just there … The *sameness* in that'. For Pam in Yorkshire, the spaces, textures and objects of what remained of the farmhouse allowed her a sense of connection with past residents. She touched the wooden beams:

> I stroke the beams because I love the fact that they've been there for 300 years. They were possibly part of something else before that. When I'm stroking the beams [laughs] and I'm lying just looking at them, I think how many other people have looked at these beams.

This imagined sense of looking and touching over the life course of the home was enhanced further by an imagined sharing of a more emotional, aesthetic and sensual response to the natural materials – the wood and stone – forming the structural framework of the farmhouse. Past residents, Pam claimed, would have *appreciated* the beams in the same way that she did; they would not have been 'too poor to think about beauty', she argued: 'I don't buy that. I think they would have understood the beauty of the house just as much as I understand it'. And, having renovated the farmhouse, she also hoped it would still be *recognisable* to them:

> I think, I hope, they would be able to walk into that end of the house and recognise it, it would feel familiar to them. I want there to be continuity with what I've seen with what they would've seen.

The materials are given meaning through the imagined response to them by past others, inert matter forming the strongest links of continuity to changing webs of social relations (DeSilvey 2007; Miller 2008, 2010).

Pam's love for the old wooden beams reflected broader responses to embodied and visual contact with original materials, in particular acts of stroking (beams), touching (stone walls, banisters, door knobs), or looking through windows (old trees). These are mainly benign, functional and relatively unchanging objects, allowing for a safely abstract sense of time's passage – it's *nows* and *thens* momentarily falling away in the haptic act and imagined simultaneity of many hands and eyes. These objects' 'pastness' (Holtorf 2013) – including their patina, colour, form, their natural, earthy qualities, their solidity, and the slowness of their wearing down – reinforced the idea of their uncomplicated authenticity and reassuring endurance. Pam described her 'terribly tactile … physical connection with the past', and how this related to the beams' '*innate beauty*':

> I love wood, and old oak like that, which has a deep glow to it because it's so old, but that you can still see little nodules where branches and twigs would've come out. So it still has life in it, but it's – it's something about

the *organic* nature of it, it's something about how old they are, it's something about the number of lives that have been lived underneath them … And that's just *mind-blowing* to me … that it's still *here*. And will be here in another 200 years, climate change permitting. It's about longevity, connection … The wood grew, was alive. I'm kind of almost *absorbing* a sense of rootedness through it, I suppose.

The wood's time depth is enhanced by the fact that the beams, likely to have been recycled from ships' timbers, predated the house, thus relating it back to 'several more generations of humans'. Touching the beams which held up the home and emerged through its intimate spaces, connected her with faraway oceans, places and times as well as with previous residents.

Connecting to earlier residents

For Ben in south London, it is the original staircase, banister and back door handle in his home which created this frisson of connection, triggering both a sense of bodily intimacy with previous residents, but also conjuring an imagined sense of their lives:

> You do kind of think: how many people have come down the stairs? When have they come down the stairs? … The banister – because it's so tactile I often think – how many people have been here? Who were they? What were they doing? Going up, going down… It's one of the most used spaces in the house, isn't it? … It will just occur to me, and I will think, 'Oh *gosh*'.

The sense of continuity of habitation is again given a positive value, a sense of the lineage of the home: 'It's a *nice* feeling. Of permanence – of being part of the *world*, I suppose. That other people have been here, and now *I'm* here, and other people *will* be here'.

This connection to home is, then, reinforced by the connection to *others* – to strangers – at the same place, through time. Ben also reflected on the original brass handles on the patio doors into the garden: 'This is original. I love this in a way … Somebody's hand, many people's hands, have opened that, and gone out. And kids would have run in and out … quite *tactile* I suppose'.

Many hands have touched the handle as part of ordinary, repeated activities – going in and out of the garden. The original object's function and form is transparent. There is no question of the door handle's purpose or how it has been used over generations of residents, all a 'part of the house'. Chris in the West of England illustrated this connection through the imagined *in situ* sharing of space:

> I think, 'somebody else has sat at this window, just like I have. Somebody else has sat right in this fire just like I'm doing now'. It's the contact with

the past, the sense of continuity. You're part of a much longer – and it sort of puts things in context.

For those who could trace the configurations of particular rooms, further similarities of aspect could be imagined. Susan in south London had researched the dates the houses in the street had been built, working out what would have been seen out of the bedroom windows by the earlier women of the house:

> The Victorian houses [opposite] were built at the time she [the first woman] started to build this. But by the time she finished it – those ones would be there because they built them in sort of batches. ... By the time [the bed-ridden woman] got here they were all built. *That* would have been *her* view.

The view for the woman in the photograph, sitting up in her bed, would have been the same as Susan's view, sitting up in her own bed. This led her to reflect further on a downstairs room:

> This clearly is how this room would have been, isn't it? I mean there's not anything you can do to this room really that will have changed this materially from when they all lived here. ... I mean it's the same purpose, you know, and the same outlook. So this is very – this is very similar. They would have looked out and seen people coming up the path, same as us.

Like Yolanda taking tea up the stairs to the bedroom, these connections are described in relation to the repetition of *everyday* experience: looking in the same direction at the same outlook, doing the same things in the same place; a sharing of views embedded in the sharing of routines. And as Yolanda's embodied sense of relatedness with the first woman of her home suggested, it can be the *immersion* in such acts which creates the most immediate connection. Josie in the West of England offered this example:

> I remember sitting up in bed when I'd had my second child. And we were in the first floor bedroom ... and because there's a fireplace and three doors on one wall, there's really only one place where the bed can go. I remember sitting there feeding this baby thinking, 'I wonder how many people have been born or died in this space – with this *view*' ... I mean, really, that's the principal bedroom, isn't it? ... And there's just nowhere else you could put the bed than looking straight out of the window ... You couldn't really have the bed anywhere else, could you? My waters broke in here [laughs].

This, then, is a benign sense of continuity, intensified by intimate physical acts and events in *exactly* the same spot – here the feeding of a new baby. Unlike for others, the thought did not elicit an uncanny, unnerving or claustrophobic response. For most, these reflections on the repetition over time of daily life in situ were kept benign with a focus on the broader, more generic, life events assumed to be most likely to be shared. Describing the old schools in her neighbourhood, Susan reflected on the 'conveyor belt' of 'layers of experience':

> Everybody who lives here, you know … went to [lists local schools]. The people who lived here all had families and brought up their children. It's like a little conveyor belt almost, but like a really slow conveyor belt – of people. You do all your things, bring up your family, and someone else is going to come in this house in the future and bring up their family.

Yolanda described this as a 'comfortable continuum', adding: 'It's connections I'm seeing. I haven't thought about the differences really'.

Feelings, women, families: generic connections

Like Pam's belief that past residents shared her understanding of the 'beauty of the house', Yolanda preferred to dwell on the similarities with past residents, suggesting her imaginative engagement with past lives was tethered to the seeking out of the uncomplicated and familiar.[1] She added: 'Obviously it's only the bits that I can imagine were the same as mine'. Sara in her maisonette, as described, also makes this assumption of similarities between people over time:

> Obviously human nature doesn't change that much, so it wasn't all completely different. The situation, the circumstances, were different. But yes. Yes. They had the same sorts of feelings and aspirations and worries, and all the rest of it. Just in a different – *context*, really.

Like Penny, in emphasising the more internalised similarities of *feelings* between *now* and *then*, Sara acknowledged that this included people's 'worries'. But these were subsumed in a broad normalising narrative – of those natural 'ups and downs' and events such as births and deaths, which no one escapes. For Lillian, on the major south London road, this amounted to the 'cycle of life', which had to include the 'continuum of people moving in and enjoying things or *not* enjoying things'. Thus, again, the emphasis was on the *sameness*, on the sweep of repeated events and feelings across time, rather than on any specific individual responses. Reflections on past lives generally dwelt on the generic and on normative roles, not just for want of further details. For example, participants tended to assume that women would be more *present* within the home. Karen, as described, imagined past women being in the kitchen, dealing with the children, carrying out practical household tasks, an iteration of gendered domesticity. Yolanda reflected on this: women 'weren't really allowed to be very individual in those days'.

Sara also drew on her understanding of women 'working away all hours presumably to run the place'. It is only the 'real' presence of Gemma's ghost that allowed for a sense of the assumed sharing of individual *characteristics* – the women ran their own businesses, did not depend on men, and were strong, independent 'spirits', thus escaping the bonds of one form of gendered sameness whilst reiterating another.

Also ubiquitous is the assumption of the home as a place for *families*, and it is here that participants fall upon a normative fantasy of the 'happy family' ideal, one which they believed contributed centrally to their homes' happy atmosphere. Ben remarked on the bad atmosphere of his house when he first viewed it, complaining: 'It didn't feel like sort of a warm *family* house', but he then reflected on whether this was a modern tendency:

> I don't know if it's true, but the nature of how we emote about houses, and see them as being *family* houses – whether this is a ... recent thing, or whether people have always done that, I don't know ... Did they feel the same way about their homes as we do?

Laura, living in a converted brewery in east London, expressed relief in discovering that the head brewer lived in the building with his family: 'It's just nice to know that people had families and *lived* here' – a positive connection in the context of her response to an otherwise 'unhomely' social context. The desire to see their own, often heteronormative, family lives reflected in the lives of past residents also influenced responses to the census records. Karen in north London imagined 'little happy families' in her home; the census records appeared to confirm this. Harvey and Rochelle in their Georgian house in central London described how they preferred to dwell on past residents during the earlier eras, when it was a more 'stable' family home – before it was split into flats and a brothel. Susan also described how she wasn't interested in finding out about an earlier resident who did *not* have a family as she couldn't *relate* to him as much as the others, and Yolanda made a similar remark about the most recent residents. Karen, meanwhile, got particularly excited when she discovered a similar family, pointing to the fact that the children were the same age as her own at the time of the census – four years and two years. And Josie also described being 'quite heartened' by the family in the 1891 census:

> And they've got two boys and two girls and you sort of think, 'Yes, okay, that's nice to think of a family' – not quite, not *similar* to ours but in a way most similar to ours so far. ... The first census that really has a sort of *proper family unit* most similar to ours.

In contrast, she reflected on a previous use of the house as a girls' boarding school: 'You think probably [they] weren't all that *happy*'. The family in 1891 represented a *happy* time for the house; the boarding school a time of an inadequate substitute family, possibly a space of loneliness or lack of intimate care.

Participants who did *not* have children reflected on wider family histories to make connections with past lives. Dan, who lived with his boyfriend in east London, exclaimed as he read the 1901 census: 'Eight alive and two dead – so ten children. My *grandmother* had ten children. She was an Indian in Uganda. Half survived. Half didn't'. Meanwhile, couples living in the larger houses who had not had children tended to have the most complicated relationship to their home's identity – built originally, they assumed, with large families in mind. The sale of recently-divorced Leah and Alex's house was coloured by a sense of lack, reflected in a desire to offer what they had failed to give to it – a *family* to fill the large space properly. Alex had spent decades immaculately restoring the house. His first wife had died young, before having children, leaving him alone. Leah – who rented a room in the house next door when they met – recalled being invited inside:

> I thought, 'This is lovely'. But I also thought it was a bit sad because ... there were loads of empty rooms with no furniture in, and stuff like that. And I thought, this is a bit of a strange house, really. For me it's important that – it's a *family* moving in here. Because this house doesn't have any kids in it – for years. It's important to me because we couldn't have kids ourselves. I feel – guilty about living here and not using all the space. So it's nice to know that this – this place is going to be a *happy* house ... a family can be happy.

Connecting through coincidence

Finding familial connections with past residents through the census records was a strategy for enhancing personal feelings of belonging to participants' homes. Another common narrative was to suggest that the home itself had called out to them to belong in it. Statements such as: 'I walked in the front door and I loved this house. I don't know why', 'it was meant to be', and 'I was fated to buy the flat' dwelt on coincidental connections and heightened feelings of coming 'home', of being *granted* a place in the home's unfolding story. Prior ideas about a home which prove to be true were also common. Jack in Bristol discovered a shoemaker called Alfred had lived and worked at his house for over 30 years: 'There's a shoemaker further down the road. As I walked passed I thought "this is exactly who would be happy here" ... It's really strange that, such a coincidence. I must have had a feeling'. Other coincidences involved discovering local family connections. Alice, who bought a house in a rough part of east London frowned upon by her upper-middle-class family, discovered a family member had lived on a parallel street: 'That was an extraordinary coincidence. You can almost see [his house] from here'. Likewise Karen's father, undertaking family research, discovered she had ancestors living locally – a 'weird coincidence' which reinforced her feeling that she had moved to the right area: 'It feels like I must have been *drawn* here or something. Somehow. I don't know why'.

Sometimes there were coincidences involved in finding out about a past resident. Cathy and her husband Jon described discovering that a famous photographer had lived in their home during the 1950s, along with, briefly, a friend of his – an equally-famous painter. Visiting an art exhibition, Cathy had been looking at a photograph of an interior shot of a home, showing distinctive wallpaper by the famous artist. Under the photograph was the address – it was her *own* address: 'So you sort of think, "What the *heck's* this?"' A few months later some people turned up at the doorstep, explaining they had been friends of the photographer: 'I took them in and gave them a guided tour, and they were very interested. They were very nice people ... But it was extremely spooky. Quite *surreal*'.

Irish photographer Julia also described a couple of 'coincidences' in relation to her famous previous resident, the poet Keats. In the process of buying the house, she had been parking the car outside at the very moment a poem by Keats was read on the radio:

> It was very strange. It was a lovely summer's evening. I sat outside in the car and listened to the poem, you know. And just thought, 'That is lovely. That is really lovely'. And then, when I moved in I [set up] my darkroom. And one of the things you do in the darkroom when you're working is you listen to the radio ... And the very first evening that I was actually working in the darkroom – what was on the radio? A play about *Keats*. It was just such a strange coincidence ... It was just a lovely play actually. And I just thought, 'Wow that is just amazing. That is just so weird. That is such a weird coincidence, you know'.

These coincidences were interpreted in relation to her growing fascination with the poet, in turn signalling the possibility of a positive connection with her new home: 'I just felt that this was a good place for me'.

Past residents as extended or borrowed family

Some participants connected to past residents by considering them to be *like* an extended family. Leah, for example, reflected on the way in which sharing the same home over time created a familial-like bond which made past residents akin to 'relatives because they shared this house. They are almost like your extended family because – they shared this *space*. So they are *like* a family in a way'.

Karen suggested there was actually more at stake in finding out about the past residents of her home because she had made a conscious *choice* to live in the same house. They were like *in-laws*, she joked:

> It's funny. It's just another house. But it's *my* house ... The house almost feels as close as [pause] – maybe because it's *mine* ... because me and my husband own it, it's, it's very much *ours*. And these are kind of my – rather

than my family – you know, we *came* to this house. We *chose* it. That was me who did that. So almost like relatives to marry into [laughs]. You know, I chose my husband. It feels like I've *chosen* these people in a way … I guess I feel like – we have a connection because *they* chose this house too … We all, at some point, said: 'Yes, this is where we are signing up to live. This will do us, thank you' … They are accidental connections, like in-laws are. You choose somebody else and you end up with them [laughs].

For some participants, past residents become a form of family substitute, due, for example, to being estranged from real family or bad childhood experiences. For William, exploring the social history of home felt safer than family history. He described his 'aggressive' family, both his father and grandfather physically violent towards him as young gay man. House research became a 'diversion', he said. The 'family' of the home's past were strangers – that was the point:

I slip into another world. It's making connections [with the house's past] because it's safer than real people. … They're not *my* ghosts. … It's other people who are *not* connected with me. … Those from the [home's] past are not going to hurt me.

For Leah, who, as described, was about to leave the home she had shared with ex-husband Alex, conducting house history research had been a way of 'cementing' her own connection to the home that her husband had shared with his first wife – a presence in the house she needed to deal with. She joked that she was the 'second Mrs de Winter', relating her situation to Daphne du Maurier's novel *Rebecca* about a young bride attempting to step into the shoes of her husband's dead first wife. Leah had produced a detailed folder with information on past residents, and within this she included photographs of both herself and of Alex's first wife. By acknowledging this woman's time in the house but also her *own* time – externalised in an organised folder of words and photographs – she was able to formally take her place within the extended family of the home. But this had also become part of the process of moving on after the divorce and house sale:

I feel I know all the ghosts of the past. And I'm one of them as well now. So I'm like, 'Oh my spirit is always going to be in this house'. Because these people's spirits are still here … they lived and breathed in this house and had a life in it. So did I. And you can never take that away … I'm cementing myself in it, I suppose.

The process of research allowed her to see her own place in the genealogical pantheon. She consoled herself that she had earned her place in its history.

Elsewhere, interest in past lives formed part of a process of creating *new* family. Susan described how she had not seen her family since she was 19, having decided to cut herself off from a 'violent' upbringing on a north London council estate:

> I'm interested in looking at the history of my home because ... I come from a very dysfunctional family ... They weren't very nice people. I had to move myself away ... It was violent. It wasn't a very nice family to grow up in ... And partly the fact that it would be difficult to start doing any family history because I've got no access to anything – photographs, stories, folk memories, whatever. But also because I don't have a connection. Because my family are *here*. The family I've *made* ... It's not obviously an attempt to gain a different type of family, but there's something – the connection with them is that they've lived where I'm living now.

Elsewhere, this sense of the 'family' of home was expressed as a desire for vicarious rootedness. Pam in Yorkshire had been brought up as a child by her mother – a 'nomadic' existence. She said:

> I've lost count of how many times I've moved in my life ... And I don't really come from *anywhere*, so I don't have a geographical connection. And the only time in my life I've really felt one is *now*. A sense of permanence.

Older houses offered her the sense of 'an anchor to the earth' she craved, as she explained:

> It is tying me to the world in relation to both place and people. [Local] families ... can trace their origins back to the Doomsday book. And ... because I can't – I don't know anything about my family particularly – I find that *intoxicating*. Scary, but intoxicating. The idea that you not only have a connection to family, but a connection to *place* that goes that far back in history.

Awareness of socio-economic difference

Pam spoke of her 'desire to share in that sense of belonging'. But after nine years in the dale – having moved there alone after the death of her long-term partner – she was also aware of the more fraught, uneasy socio-economic context, a 'discomfort' about being part of the trend for people to leave the city and buy into an idea of the idyllic rural dream:

> I'm conscious that the actual local people would be very cynical and dismissive about that ... It's a cliché, you know, you're only regarded as coming from here when you've been here 300-plus years. There's a lot of resentment about people like me, that we don't have that rooted connect-ivity and that we can buy our way in that they can't. So I'm conscious of both those sets of feelings ... of a discomfort that I'm one of those

people that is contributing to the fact that the *actual* people who do come from here can't afford to carry on living here ... The only reason I can be here is that, because of the vagaries of the property market, I had enough money to do it. And that, you know, the actual sons and daughters of the dale can't. It's only incomers, offcomers like me who would have done it.

These complicated feelings contrasted with the sense of rootedness that encountering her home's original features afforded her. She admitted ruefully: 'Stone and wood don't come loaded with any guilt, do they, about how much easier one's life is now'.

Pam's discomfort is compounded by her sense of how hard previous lives would have been. Contradicting her belief that previous residents would have taken the time to appreciate the beauty of the beams, she argued that their lives were about mere 'survival': 'You didn't really enjoy, have much time and energy to sit around on your bottom and just talk'. The dale's exposed environment was known to be particularly harsh. There had been an exodus from the valley during the twentieth century, with only one farmer remaining to brave the elements; Pam's attention was drawn to the contrasts of life lived in the home, a further way to reflect on her sense of privilege. Against this, her working-class background allowed her to justify a sense of connection – 'and a pride in that connection as well. I love that connection. But I'm also uncomfortable with it'. The values she admired the most – functionality, humility, restraint, adding nothing extraneous – are, she admitted, no longer represented in the farmhouse after her renovation: connecting the house to the adjacent barn had doubled the floor space: 'You've got a house that had 14, 15 occupants at the end of the 19th century, and now I've got twice the space and I live in it on my own'.

Number of residents

Pam's uneasiness about the differences between *then* and *now* focused on basic numerical data: the number of people living in the house. A cursory count of residents listed in the census records was enough to offer participants an imagined sense of how life was lived differently in the past. This became a key expression of feelings of economic privilege:

'Gosh that's crowded isn't it? Look at the age range'; 'All in here? I mean surely they couldn't fit 2,4,6,8,10–*12* people? No they couldn't'; '*This* number of people [laughs]. There are quite a lot of people packed into this house. ... In so many ways I am very fortunate, and really quite privileged'.

This contrast was a particular theme in London with its Victorian terraced houses built for the working classes but now too expensive for many people. Julia described how in the first census there was about eight people living here, one of whom was a coachman ... so it was incredibly crowded, right? Eight people! I mean I live here on my *own* [laughs] ... I just think it's amazingly luxurious [laughs] ... The difference is quite striking when you look at the records ... This [street] was pretty down-market ... I just feel so lucky'.

But she added defensively: 'I mean it's not a huge house, you know. It's only a small house'. Karen also reacted to seeing, in one census, eight children listed as living in her house; she herself had decided there was no room for her to have a third child. But this led to a counter-view: as a *middle-class* woman, why couldn't she afford properties originally built for *her* social class? She elaborated:

I feel *aggrieved* that – people with the education that my husband and I have, and the careers that my husband and I have, lived in a damn sight bigger house when this house was built. This was a – a *poor* person's house. Whereas now you need an *enormous* salary to afford the mortgage on one of these houses. So I think that also is funny. That, you know, what drives me mad having such a narrow hallway – whereas middle class people like us, without sounding snobby, would have had a much bigger hallway when these houses – you know, these houses were never built with the idea that university graduates with good jobs would live in them [laughs]. So that's the other way round ... I just think it's interesting the way things change over time. Sounds very snobbish to say that, but it's true you know.

Pointing to the mansion in the local park – now a café with meeting rooms – she explained that a banker once lived there, but now a banker (like her husband) had to live in a small terraced house built for the uneducated poor. Karen's responses to the contemporary impact of socio-economic change focused on her own expectations and entitlement which she felt the current London housing market had curtailed. Others also responded to the changing socio-economic status of their neighbourhood, including Harvey and Rochelle, the first owner-occupiers of their Georgian house. Harvey explained:

The home was owned by landlords before. It didn't have drains until 1903 when the council compulsorily drained it ... The *contrast*. I mean, we think this house is too small for two people. And yet it had 14 people in it in the 1871 census, you see [laughs]. It's kind of a – *sobering* thought to realise how much things have changed in such a short space of time. And how much better we live than they did ... I don't feel *guilty* about it at all. But it is – I guess, I guess I feel kind of *sad* for them because they didn't have the same things we've got ... And it must have been tough.

The precariousness of past lives was also demonstrated in the census records. Sara, for example, reacted to the stark listings of the 1911 census which showed that three out of six children had died:

I mean you *know* this … You *know* there's a huge mortality rate. But then you sort of *see* – she had six and – only *three* survived. It's [exclaims] eeow! [Deep intake of breath] … There's something really quite poignant about all this, isn't there? You know, the little children, the *what* they did, the deaths. This [young man] probably went off into World War One, didn't he?

Degrees of comfort at home

Those participants who lived in homes originally housing working people assumed that past lives would have been far more difficult, particularly in relation to issues of personal comfort and space, cleanliness and domestic work, and sensual deprivation. Yolanda, for example, reflected on the domestic life of the first woman:

> It's also the thought that whatever she did in her life it would have been a lot harder for her. Because her life was probably a lot more *physical* than mine – hand washing socks and children's shirts and things – no washing machine, no tumble dryer.

Speculating on the high number of people in her home, Julia's thoughts also turned to issues of hygiene:

> They must have all been sharing rooms and everything … It's really hard to credit. I mean there are four rooms – one, two, three, four, right – downstairs and the back extension was built later … There would have been a tiny scullery out the back and … if people did what they called 'nights', that went into a kind of a tank at the bottom of the stairs, right, and was emptied. So the place would have *reeked*.

Leah and Alex, their house adjacent to a now-abandoned railway line, reflected on the grime from the steam-trains – the permeability of boundaries meaning an inability to escape from the fine airborne dirt. Alex said:

> You have to remember back in those days it was *steam* … We have a steam train come through here occasionally on a Saturday morning, one of these, you know, *specials*, yeah? And I tell you, the noise and the dirt coming off of that. Now if you have that every 20 minutes, as you would have done in the old days, you know … You got soot round the doors. Everywhere. You know. It gets in the door jams, got everywhere, you know. So it wouldn't have been a very pleasant place to live in the steam era.

Others point to more recent forms of local deprivation. Cathy and Jon described how the famous photographer who lived in their house took photographs from the front step as a form of social commentary on East End life. When showing around his family and friends, they were surprised to see that the house backed onto a canal at the bottom of the garden – now a prized feature. Why had he not taken any photographs of that?

CATHY: You'd think a photographer – I mean, he was interested in people and techniques and things – but in the 50s, the canals were just filthy, unpleasant places … pretty stinky. Jon's dad got a certificate of bravery for diving into the canal to save someone from drowning in 1932. That's where he had his bath.

JON: Because there was the bath house, a. you had to pay, and b. it was often over-subscribed. So my grandfather used to take him into the canal with the soap.

The history of the canal – 'filthy' despite being used to bathe local children – was of no interest to the photographer, whose 'mother was part of the Bloomsbury set', Jon explained. He considered himself to be 'slumming it' by living in such a place, which gave him a vantage place to record the social world of the lower classes. But he ignored the canal in his back yard. Cathy explained:

> We deduce that the canal was a *stigma*, a no-go zone. It was dirty, grubby … It wasn't a sexy thing at all. And given he was from the upper classes, he probably didn't want to be [associated] with that kind of thing.

Not nostalgia

The focus on deprivation was also a way in which participants could prove that their fascination for the past was not merely 'nostalgia'. Gemma pointed this out in relation to her love of all things Victorian:

> It's not nostalgia. You know, I don't want to have rickets or malaria or anything [laughs]. I've got a friend who says to me – because I'm badly asthmatic, he's diabetic – he said, 'Neither you or I would be here anyway' if we were in Victorian times [laughs]. We'd both be dead.

Pam was also aware of the dangers of a nostalgic view of local rural lives, arguing:

> We are in danger of romanticising their lives. Because it must have been bloody hard. Because a. there would have been no personal space … They would have to get water from the beck, in weather like *this*. Nowhere to wash, nowhere to have sex privately … They moved the school over the hill … a two hour walk up and over, through the bog for the kids to go to school. So education was hard work. Everything was hard work … cold, wet, itchy, hungry.

Some, however, parodied their own nostalgic tendencies. Laura, in her converted brewery flat, laughed at her enthusiasm for 50s retro vintage, admitting: 'in some ways that's an escapist thing, thinking, "Oh life was always better in the 50s! The halcyon days! We're living in terrible times now!"' As described, Leah also ironically reflected on her costume-drama fantasy of the past. This led her, a little reluctantly, to stress her understanding of the difficulties of past lives:

> It was probably worse then ... I don't want to think about all that! I don't want to think about all that! [Laughs] It would have been *dirty* and *smelly* and like – yeah, *dirty*. And in here they probably would have had really *dark* walls, dark wood, green – like dark *greens* on the walls because it was easier to keep clean that way, if it was all *dark*. The whole place would have been *dark* and *smelly* with lots of – kids running around [laughs].

If Leah's desire was for an enjoyable fantasy about the past which stripped away all the social realities, she also needed to emphasise her *knowledge* of these realities in order to show her ability to juggle these two co-existing imaginaries.

Servants

A further way in which participants acknowledged the differences between their life in their home and those of past residents was to consider past social discrepancies *within* homes. For those living in the larger houses, this led to a concern for the servants. Jacob in the South of England wanted to know 'what were the notions of ... *privacy*' if they shared the 'only staircase'. Many imagined the servants living and working in places with little or no natural light, such as basements. Leah described the servants' quarters 'down here':

> So the kitchen would have been here, they would probably have eaten here, the servants. And they would have slept, several of them, in that room ... There is no natural light to this room. The servants would have slept in here.

Ben in south London had been given a book to read on the Edwardians. In it, by sheer coincidence, he had found an advert related to his *own* home:

> And when I was flicking through this and reading it, I came across an advert for a house that – I *live* in – an advert for a servant ... [Quoting book] 'Advert in local newspapers confirm the respectable servant-keeping character of the area: "Servant, £14, good cook, general or useful help wanted ... cooking and housework, £16–18. Apply [Ben's address]"'.

The advert triggered his interest about who was employed:

> Who was this person? What was her name? What was she doing? And
> how did she live? ... She would have lived in one of the top rooms. Where
> did she come from? Where was *her* family, because they were all single
> women, weren't they? You know, who had quite *difficult* lives. You know,
> being in domestic service.

Ben's sympathy for the servants' lives focused on the kitchen, which had not
been changed when he moved in. The kitchen had a scullery, door to an outside toilet and coal scuttle and – what particularly upset him – 'tiny windows'
which afforded no view onto the garden:

> It was really dark with those high windows. And every time you came into
> the room, you were just dying to look at the garden, because the garden
> was a – *feature*. But of course they weren't meant to enjoy the garden.
> Because they were working ... And they weren't meant to sit down. So
> there is enough light in here obviously for them to do washing I suppose,
> because it would have been where they did the washing. Of course it was
> a very mean space as well, because the servants would have been cooking
> and cleaning here as well, it was the servants' space. So [laughs] they
> couldn't enjoy it.

His previous image of the happy family scenario, with the children running in
and out of the garden through the doors from the living room, was tempered
by his understanding of the contrasting life of the lack of enjoyment for the
servant underclass, the house designed to ignore their comfort as they worked,
including an inability to look out onto the garden.

The past was darker

The lack of natural light described by both Leah and Ben in relation to the
lives of the servants created a further contrast with the lightness of the family
spaces. But also, the past is assumed to be generally a darker 'place' – beyond
the use, for example, of dark furnishings to hide the dirt, as described by
Leah. The dark was associated specifically with a lack of electric light as well
as used to signal disapproval of the architectural details of homes built in
particular eras. For example, Yolanda's love of 1930s buildings related to 'the
light. ... You do get more light in 1930s buildings', whereas Julia described her
dislike of Victorian houses:

> I hate Victorian. I mean that kind of fussy, narrow, scrunched up feeling
> I don't like. Sorry [laughs]. I have a friend living around the corner ... and
> she lives in this house and it's a lovely house in many ways but it's so *dark*.
> It's that period ... I would go out of my mind. I'd be so depressed because

it's so dark … It's about the light and the sight lines in the house and the way the rooms work.

Sara assumed her home would have been dark in its earlier days, and like others, she placed a negative value on this (it would have been 'quite dark' and 'not very nice'). Jack in Bristol likewise assumed the house would have been 'quite pokey and dark', and that previous residents would have gone to bed earlier because of this. Bella and Sylvia near Bristol described finding blocked up windows at the front of their large mansion, reflecting on the difference this made to the quality of light – a source of both pride and regret:

SYLVIA: There's absolutely no doubt about it, having the new windows put in, you know, that were blocked, has made a huge difference. Yes. That's why my mother-in-law – I felt so sorry when she said: 'I can't bear it to think that those windows were there all that time and I never realised and how dark it was'. But you *don't* realise. You live with a place and you don't always see what other people see.

BELLA: She said to you, didn't she, 'I wish I'd known how much light we could have had' … I do feel sad because she had to put up with this horrible dark, unnecessary dark. Because I live for the light and I would, when I go and look at houses now, whenever I've moved anywhere, I've looked to see how much light we get.

Only Miriam, in her central London council flat, found the idea of living in lamp-light atmospheric and romantic:

I keep thinking how much more – before everything was brilliantly illuminated by electric light – for example, interiors where you just get – a *lamp*, you get lots of shadows. It's a different quality of light than if you have strip lights everywhere, which absolutely illuminate everything, and takes away the shape of things.

Length of stay

A last assumption made by many participants was to positively value the longevity of residence – over and above a consideration of particular changing circumstances – equating this with senses of stability, security and belonging. Although many participants had been settled in their current homes for many years, most had moved a number of times during their own lives. Those living in the smaller towns assumed that their lives were less transient than those living in the larger cities – a fact perhaps borne out by the number of previous residents still found living locally. Longevity in a home is also taken as an indicator of a positive atmosphere – a reassuring feature, related to the idea that stable, happy families have lived there. Josie in the West of England commented that the fact that only five families had lived in their 200-year-old house 'says something about the house really ... the fact that people get here and they stay here'. Chris added: 'Whether that's because they loved it, I don't know [laughs]'. 'I suppose I like that idea', Josie replied.

The opposite was also assumed. The more transitory a home's population over time, the more it signified economic hardship, or a harsh local environment – such as the noise and dirt from the local railway line Alex described. Sara, however, wanted to put a positive spin on the frequent moves of past residents of her home:

> So in ten years there was nobody from that lot who were still around in the house? Whether they'd gone to bigger and better things, some of them had died or ... you're doing a moonlight flit because you can't pay the rent ... Hopefully they were bettering themselves. Hopefully they were rising up in economic terms.

In turn, there were also assumptions about the relationship between *ownership*, length of stay and senses of belonging. Susan reflected on how renting used to be the 'norm' for the wealthy, 'the vast majority of people rented then'. As the legal ownership of homes was a 'relatively recent thing', there was no stigma attached to renting, Ben added, and it did not preclude people from staying a long time in the same home. He reflected on his family experience:

> My parents only owned their house from 1970. Prior to that they had only lived in rented accommodation and their families rented – etcetera. It was true for the majority of the population. You were very, very lucky or rich to own a house. ... And they had fewer *possessions*.

But remarking on the high level of transience in the records, Karen reflected how these did not suggest the 'little happy families' that she had hoped: 'None of these censuses have any consistency – *at all*. I mean, there is no consistency. ... There is no link to – the idea of settled families'. But she played down the negative implications of this by relating it to how people she knew moved home, assuming that this had always been true: 'That's how this area is a lot at the moment. You know, people moving, have the baby, stay a few years, move to the countryside, vacating it for somebody else. ... People move around a fair amount'.

Hidden queer histories

These assumptions about the lives of earlier residents, reflected through a mix of knowledge, imagination and conjecture, suggest how participants actively engaged in reflecting on the similarities and differences between their own and past lives in their homes. Whereas it was far easier to enjoy the idea of the similarities and connections, the differences also needed to be dealt with. In general these suggest that, rather than an uncomplicated sense of nostalgia for a 'better' past, people tended to emphasise – and in general wished to empathise with – past lives as more *difficult*, particularly because many participants lived in homes which had been built for the less wealthy or, for those in the larger homes, had housed servants. Nonetheless, compared to the more complex task of encountering *living* people – as will be explored in the next chapter – *earlier* residents could be imagined at a relatively safe distance and were generally subject to normative stereotyping. Most found comfort in an assumption that theirs were homes of happy, stable heteronormative families, or sought out examples of past lives most similar to themselves. This might suggest a reluctance to acknowledge particular differences of identity or experience. Indeed, the discussion of differences (beyond observations that participants might have had little in common with past residents as individuals) tended to focus on the more generic and broader observations related to how life was lived in the past.

Even when the motive is clearly to be respectful in not making assumptions about past lives that cannot be known, the avoidance of dwelling on potential differences can inadvertently leave certain aspects of these lives hidden. To end this chapter, I will offer an example of a previous resident who, the narrative appeared to suggest, might have been a lesbian. The *not naming* of this as a possible aspect of identity resulted in a story which stopped short. Here a *lack* of speculation about the woman's life left a silence akin to a refusal to engage. Whereas elsewhere the gap between what was knowable and unknowable about past lives was considered with frustration, anxiety or relief, here a lack of acknowledgement of the gap itself led to a particular affective fissure.

Chris and Josie in the small West of England town described how the previous owner of their house, a Miss H, had lived in the house since a child in the 1920s, inheriting it after her parents' death. She had been living for two years in a retirement home by the time the house was sold and appeared to die at a crucial moment. Josie jokingly described her wilfulness beyond the grave in delaying the sale, the mixing of tenses here suggesting her presence was still felt: 'And we never met her. She died, *typically of her*, the day we were due to exchange contracts. And delayed it by three months. She's quite a character the lady that was here before'.

Their account of Miss H reflected their fascination, affection and understanding of her based on what she had left behind – a material inheritance which included 'little bodgey' DIY solutions around the house, such as a cat flap made with the end of a glass milk bottle, still used (Josie: 'We just know that's Miss H ... This is why we like her [laughs]'). She had no interest in modernising the house; there had been an outside toilet and wood-burning fires for heating:

> To stop it freezing up in winter was an electric light bulb inside a baked bean can with a long trailing cable all through the house to a little two-point socket in the dining room. And that was her way of sort of stopping the pipes freezing, to leave the light on all night [laughs].

They also knew Miss H 'smoked like a chimney': the walls of the downstairs room she made her home toward the end of her life were stained brown. Josie's father had 'washed the walls down' leaving 'his arms stained brown for weeks … He couldn't get it off'.

Miss H also took in injured animals and left all her money to an animal charity. The couple dug up 'all sort of animals' in the garden and noted that 'everything was all sort of scratched and clawed' on the walls of an upper room where she had kept birds, leaving the window open 'so they could fly in and out' (Josie).

These traces of Miss H in the house lent themselves to their idea of her as an archetypal 'eccentric spinster'. But further specific clues to her identity emerged out of a description of her physical appearance. They listed what they had gleaned from neighbours:

JOSIE:　Never married but she's supposed to have been quite a character. Rode a motorbike. Wore a shirt and tie. Smoked like a chimney. Eccentric animal lover.

CHRIS:　One of the first women in the [town] to ride a motorcycle … She was driving a motorbike during the War.

JOSIE:　We were told she wore a shirt and tie and had her hair cut very short … She dressed like a man, she wore a shirt, collar and tie … A shirt and tie with burn holes down her shirt … *Quite* a character.

The description of this woman who 'dressed like a man' was not elaborated, and they appeared to prefer not to speculate further, closing down any detailed discussion about her life. When asked about her masculine appearance, Chris emphasised defensively: 'In *old* age … In the last years of her life. That's how the neighbours remember her'. Asked if she might have had 'partners', Josie deflected the answer onto a comment on the house: 'I don't think anybody else would live in here', and the conversation moved on. If they believed Miss H might have been a lesbian, they were reluctant to voice it. This could well have been a protective response, not wanting to assume or speculate about her sexuality – the element of her identity they could never know for certain. And they were keen in other ways to maintain a respectful distance by, for example, not using her first name, deferring to neighbours' reference to her by surname, arguing that such a 'formality' reflected the age in which she lived. Josie added: 'I think we were told she'd like to be called Miss H. [Otherwise] she would be quite affronted'.

Speculating about her sexuality might, equally, be deemed disrespectful, crossing a boundary of etiquette; perhaps the 'spinster' label and description of appearance was deemed to suffice as a polite indication that didn't need spelling out. But this has consequences. Remaining affectionately but generically 'eccentric' might bleach out aspects of her identity which may be uncomfortable to speak about – perhaps even unacceptable or repugnant. But doing so closes down important possibilities for exploring wider historical contexts,

Figure 4.1 Ben described his sense of connecting with previous residents when holding onto the original banister, triggering questions about their lives.

including questions about mid-twentieth century life for gay people in small semi-rural towns. Indeed, as some queer scholars have argued, it is better to use a descriptor such as 'lesbian' even when exploring those whose self-identity is not fully explicit, or where such descriptors were not culturally available at the time. This is because the 'construction of queer existence as an "impossible object" of historical inquiry' has led to gay people being 'marginalized not only through moral censure but also through silence and disregard' (Love 2007: 133–134).[2]

Conclusion

This chapter has illustrated the importance of attending to the effects of perceived similarities and differences between past and present lives through which people negotiate feelings of connection or distance, drawing upon and reflecting their own beliefs, values and assumptions. As a counterpoint to the sometimes anxious responses to beliefs in the continuing presences of the past, outlined in Chapter 3, expressions of awe at feelings of continuity and permanence arise when reflecting on the sweeping and safely abstracted temporalities of homes. Here there is a frisson gained by reflecting on one's temporariness of residence in the face of the broad lineage of previous and future inhabitants – feelings not just reserved for the oldest buildings but enhanced by haptic encounters with older features. The sense of being part of something bigger, and of feelings of continuity, becomes a pleasurable outcome of reflecting on the home's pasts and futures.

In general, as described, people seek to enhance connections with the past as a way of reinforcing their feelings of belonging at home. This includes imagining similarities – dwelling on the repetition of routines or uses of domestic space, reinforcing social norms or generic knowledge, or assuming shared human feelings across time. Emotional connections to the house itself are also expressed through narratives of coincidence, enhancing a sense of the inevitability of being 'in place', the authenticity of one's sense of fit.

However, participants also acknowledge and contend with the differences between their lives and those lived previously in their homes, again dealing with the more specific ways in which the past's 'otherness' manifests – the broad socio-economic disparities and the ways they are deemed to manifest in different ways. The sense of relative temporariness within the long, linear sweep of time is rather different, for example, to the negative socio-economic value given to some past inhabitants' *actual* temporariness of residence. Broader imaginings of the past do not always reflect, and can be challenged by, the more detailed fragments of specific information afforded through the archival records or oral histories. Assumed differences often reinforce past hardships – crowding, lack of comfort, cleanliness or access to light (Dowd and Hensey 2016), and the limited capacity to feel at home for those *employed* within the home. Reflecting on such differences creates a range of emotional responses, including guilt, unease and defensiveness, but also feelings of empathy and care. Differences are also in part ameliorated through imaginative 'cherry-picking', compartmentalising the past's more negative elements. Forms of 'cherry-picking' can also be used to avoid reflecting on the possibility of particular forms of identity which might be uncomfortable to recognise or speak of, such as the sexuality of a past resident. But such strategies are rarely complete; participants are often conscious of the cultural ideas they bring to their imaginings and speculations about the past, including the elements of the past they wish to emphasise or avoid.

Daniel Miller is thus wrong to assert that 'feelings of alienation' (2001:107) inevitably arise through an awareness of pre-habitation, but nor, again, is it quite correct to assume that a focus on one's home's history will always positively enhance uncomplicated senses of belonging and connection (Austin, Dowdy and Miller 1997; Bushell 1989; Tindall 2006). Framing these relations in terms of 'domestic genealogies' (Lipman and Nash 2019) allows for a more nuanced sense of relatedness to past residents (quite literally for those describing past residents as forms of 'family' beyond a narrow definition of kinship), offering a context for opening up questions about which selves and others are included or excluded when facing the challenge of experiencing imaginative co-habitation in domestic space. Beyond residents from the more distant past – those who are generally least knowable and with whom communication is at most oblique – how far do people relate to the more recent former occupants of their homes and negotiate senses of belonging with these different selves and others?

Notes

1 Research on how past people are viewed by present ones has tended to focus on either the study of famous historical figures or responses to the more immediate familial dead. In both cases, a commonplace emphasis is on how the dead become identified in ways which reflect present-day concerns rather than the realities of past lives. For example, historiographical accounts of writings about famous past people suggest the ways in which 'each generation has ... left us their *own* version', including the ways that 'earlier generations ... filtered our picture of the past' (Liddington 2010 [1994]: 10). In relation to the more recent familial dead, Daniel Miller (2010: 150) argues that the deceased individual 'gradually transmutes into an ancestor by becoming largely remembered with respect to an idealized category'. For example, the figure of 'gran' is 'communicated to the grandchild ... without faults and foibles'; through the left-behind objects associated with her (the 'old clock or washing mangle'), she is turned into a 'kind of museum figure evocative as much of her period as of herself' (see Bennett 2018).
2 Alison Oram (2012: 540) has also described the way historic houses which are open to the public have often downplayed the lives of past lesbian and gay inhabitants. At Shibden Hall in Yorkshire, once home to nineteenth century lesbian diarist Anne Lister, for example, Oram discovered that Lister remained rather a 'secret', with the house's history framed around the theme of heteronormative families (a situation recently rather dramatically reversed with Lister becoming a marketable asset after the success of a BBC/HBO television series on her life). Oram (2011: 205) also suggests that historical houses' materiality and 'heterotopian nature' can reveal 'layers of non-normative domesticity', pointing to the 'complexity of alternative sexualities'. We might consider how people's encounters with forms of 'otherness' or difference experienced in *domestic* homes might hold potential to subvert (or merely replicate) the ways lesbian and gay past lives are framed within public historical settings.

References

Austin, D, Dowdy, M and Miller, J (1997): *Be Your Own House Detective*. London: BBC Books.

Bennett, J (2018): 'Narrating Family Histories: Negotiating Identity and Belonging through Tropes of Nostalgia and Authenticity'. *Current Sociology* 66.3: 449–465.

Blunt, A and Dowling, R (2006): *Home*. Abingdon: Routledge.

Bushell, P (1989): *Tracing the History of your House*. London: Pavilion Books.

DeSilvey, C (2007): 'Salvage Memory: Constelling Material Histories on a Hardscrabble Homestead'. *Cultural Geographies* 14: 401–424.

Dowd, M and Hensey, R (eds.) (2016): *The Archaeology of Darkness*. Oxford: Oxbow Books.

du Maurier, D (2004 [1938]): *Rebecca*. London: Virago Press.

Holtorf, C (2013): 'On Pastness: A Reconsideration of Materiality in Archaeological Object Authenticity'. *Anthropological Quarterly* 86: 427–444.

Love, H (2007): 'Impossible Objects: Waiting for the Revolution in *Summer Will Show*'. In L Doan and J Garrity (eds.) *Sapphic Modernities: Sexuality, Women, and National Culture.* Basingstoke: Palgrave Macmillan.

Liddington, J: (2010 [1994]): '*Presenting the Past: Anne Lister of Halifax 1791–1840*'. Hebden Bridge: Pennine Pens.

Lipman, C and Nash, C (2019): 'Domestic Genealogies: How People Relate to Those Who Once Lived in Their Homes'. *Cultural Geographies* 26.3: 273–288.

Miller, D (2001): 'Behind Closed Doors'. In: D Miller (ed.) *Home Possessions: Material Culture Behind Closed Doors*, 1–19. Oxford: Berg.

Miller, D (2008): *The Comfort of Things.* Cambridge: Polity Press.

Miller, D (2010): *Stuff.* Cambridge: Polity Press.

Oram, A (2011): 'Going on an Outing: The Historic House and Queer Public History'. *Rethinking History* 15.2: 189–207.

Oram, A (2012): 'Sexuality in Heterotopia: Time, Space and Love between Women in the Historic House'. *Women's History Review* 21.4: 533–551.

Tindall, G (2006): *The House by the Thames and the People Who Lived There.* London: Chatto & Windus.

Varley, A (2008): 'A Place Like This? Stories of Dementia, Home, and the Self'. *Environment and Planning D: Society and Space* 26: 47–67.

5 Belonging to home
Negotiating ownership

This chapter explores how senses of self and belonging to home are moulded, supported or challenged by self/other relationships which are formed through perceptions of similarity and difference, focusing on the more immediate scenario of responses to recent residents. Unlike the imagined (and assumed to be long-dead) presences of the home's earlier pasts, recent inhabitants offer possibilities for 'fleshed out' encounters and leave behind more specific material traces. Considering this more contemporary form of heritage in the home reflects forms of 'inheritance' which are often ongoing, encompassing the home's presents and futures as well as its pasts.[1]

People's encounters with these more recent residents often start with the point when one inhabitant leaves and the next enters a home – particularly at the transfer or exchange of ownership. This is where this chapter will start. It will move in turn through the different temporal moments which inform relationships between recent residents, underpinned by negotiations of belonging to home – their *continuing* presences, different forms of *return*, encounters with *future* inhabitants and imagined futures for the *home itself*.

The exchange process: blessings and curses

The moment of exchange or transfer can be a fraught or pleasant experience, and despite its brevity, can have profound effects, suggesting that the previous owners hold power to confer forms of blessing – or curses. The intensity of moving house is often vividly recalled, even after many decades. Brian and Megan, who had lived in their suburban 1950s house since the 1960s, recalled the exchange as a positive experience:

BRIAN: They told us who their doctor was. They knew Megan was expecting a baby and they told us he held maternity clinics.
MEGAN: And they told us when the dustbin's collected.
BRIAN: And they told us what decorations had been done, and what kind of paint had been used ... And she *typed it all up*.
MEGAN: And they left the house *absolutely spotlessly* clean ... You know, we didn't have to do *anything*.

BRIAN: It was that kind of practical – she was a sensible woman ... and she did things *properly* ... They'd *kept* it so nicely. They'd *cared* about it ... We liked the people we were buying from.

In Keats' house, Julia also received a written note from the vendor, which she considered a 'blessing' – smoothing the ritual of transfer. The previous owner had moved to North America and rented out the house before selling it: 'She sent me a lovely letter, which I still have, wishing me well and saying, you know, she was glad I had the house, and it's quite a *special* house – which I was so touched by ... The letter represented a blessing to me'.

Judging recent inhabitants

The letter affected Julia positively to the extent that, sixteen years later, she still kept it in the house, like a gift. For others the experience was not so positive. Sandra's description of her ex-neighbours' curse of their already-jinxed flat suggests that some parting 'gifts' might be driven by revenge or a desire to claim back power. Many participants judged previous residents in relation to their behaviour during the exchange process, but also important was the material condition of the home as it was left – the décor or surface decorations. Gemma combined a scathing description of tensions during the moving process, the fraught aftermath, and the vendor's material choices for the cottage:

> When we moved in he was rushing around with a clipboard, getting in everyone's way and ticking things off. He wouldn't pay for two small removal vans that could have come through the [adjacent] school archway. He would only pay for one large one which had to go at the end of the footpath. And he had two spaniels in a cage. And if it had been me I would have put him in the cage and let the dogs out. It was gone twelve o'clock. A friend of ours – who looks like a thug but is actually a social worker – said, 'We want to move the stuff in'. And the man said, 'Well you can't because we haven't moved our stuff out'. And [my friend] looked at his watch and said, 'Hmm. It's gone twelve. This piano's worth twelve thousand quid. It's not standing out here on the patio'. And that was it. We started moving in [laughs].

After the move, she added, the vendor

> kept pestering us ... wanting more money for the carpets and curtains. ... So in the end I said, 'Mr B. If you have room in your new bungalow for a 17-cornered carpet, then by all means – *take it*' [laughs]. He realised that, if you look, all the windows are different sizes, and there's no way on God's earth this carpet would fit anywhere else ... When we replaced it [after many years] the man who came to measure ... we said, 'will it be

you fitting it?' He said, 'No, I'll be off that day'. And we said, 'Well you don't know what day it will be'. He said, 'I'll make sure I'm bloody off that day' [laughter].

The humour in her anecdote revelled in the quirky shape of her cottage whilst reinforcing her view of the vendor. But she was also scathing about how she had found it:

> To alleviate [the lack of natural light in the bathroom] they put in a yellow suite, with yellow tiles and a yellow floor [laughs] ... The kitchen and bathroom had to be done because it was *horrible*. There was a stainless steel sink for a start. In an old cottage – why would you do that? It said, 'attractive pine-fronted units'. They were only pine because that was the top layer of the plywood. A cork tiled floor, but where the tiles had come up, they were nailed down, glued down, screwed down. And when Alex took all the flooring up there was about four layers of lino, and underneath that was just – *dirt*.

Other participants also complained of the inappropriateness of fittings. For Rita, this reflected her sadness at letting go of her old family home, ceding control to the whims of other people's taste: 'A lot of the personality has gone from the house ... Look, that's the fireplace that they put in which ... frankly looks like it was bought from John Lewis or something'.

Martin's similar observation allowed him to reinforce the way *he* had improved the home: 'They put in, you know, five pound chandeliers which vaguely looked as though they might just have been Victorian or something like this but quite patently were bought from Woolworths or equivalent. I wasn't going to do that'.

Judgements of character were also made in relation to décor considered in bad taste or poorly fitted. After a rather nasty tussle over the price of the house, Anna described her disdain for the polystyrene plant pots the vendor had left behind. Ben also mixed descriptions of the home as he found it with his judgement of the previous residents – the 'shoddy' house of 'odd' people:

> It was a very, very ugly, unloved house with grey pebble dash, and it was cold ... It was just very [deep intake of breath] one of those houses you walk into and get a bit of a *shiver* ... It looked very lonely and forlorn ... It had horrible wallpaper on the ceiling for some reason – it didn't feel *lived* in. And they were sort of quite *odd* people ... They had done up the house in a very, very, sort of *shoddy* way.

Elsewhere, Jack in Bristol had encountered a hostile vendor ('she was really aggressive and nasty and unfriendly'). In turn, he judged her for the 'damaged' home she had left: 'She lived here for 30 years and she allowed the house to be damaged ... There's a leak above that door and those windows, which poured water down and rotted out the floorboards down here ... I can't understand someone who's got a house and treats it like that, you know. It just doesn't make sense'.

Judgements of character, however, can be softened depending on context. Adam, for example, described his vendor as 'dodgy', illustrating it with a story of how he 'nicked the fireplaces' and only 'reluctantly left us a padlock for the garage'. In contrast, the neighbour admitted later that he had stolen the original window shutters. Adam's take on this was rather more forgiving; he legitimised it as 'salvage'. He was friendly with him – a 'very nice guy'.

Leaving things behind

Unwritten rules of etiquette governing the exchange process included the correct behaviour in relation to what one takes or leaves from a home. Fixtures were generally deemed to be a part of the house, but more complicated were those which participants themselves had added. The question of whether to leave a designer bookshelf attached to a wall, for example, led to a heated discussion between one married couple. Cathy described dilemmas over items such as fireplaces, art deco kitchen cabinet handles and a back door. The fireplace was non-negotiable – she had taken it from a previous home, commenting defensively that in any case it 'would be skipped' if left behind. She also 'couldn't bear' to leave the back door which she had put stained glass windows in, but clearly needed to replace it. She was forced to leave the handles, reassured that the new owner appreciated them: 'I couldn't have taken the handles because the modern ones are a different size for the screw holes and it would have wrecked the kitchen. It broke my heart to leave them. But the bloke that bought it really liked the kitchen, so I don't think he ripped it out'.

Less complicated was the ethics of leaving a home in good order. William described this as a moral duty. He intended to 'leave this space spotless ... It's handing over. It's what you do. You *hand over* ... Except I'm going to leave the curtain tracks – there were no curtain tracks here'. But many complained of having to clear away junk. Sometimes this was left by builders from an earlier era – such as rubbish found under floorboards at Harvey and Rochelle's house, and rubble filling up Gemma's well under the living room floor. But for Jack in Bristol, the 'whole load of rubbish' left – along with a doll's arm stuck to a wall, newspaper, bottle tops and corks, and a broken chair ('my housewarming present') – was just another personal insult. Anna's difficult moving-in process also ended with a bad-tempered exchange about the cellar 'full of old junk'; she told the vendor to 'get rid of all your old rubbish'.

But for tenants the situation was often more complicated; there was less control over what others left behind. Sandra also had to deal with an inheritance of material neglect: 'My first impression, when I first saw it was, "what a dump" ... They even took the loo seat. This place was not really fit to let ... it was in dire need of decoration. There's actually a hole in that ceiling. They promised a lot of things, not all of which they did'.

Continuing presences

Despite the fact that the most recent residents are known or are assumed to be still alive, for some their lingering presence is still *felt* and needs to be dealt with. Laura in the converted brewery, for example, had bought the flat off a bank, and was concerned to hear that the previous owner had gone bankrupt after a business failure:

> The story did affect me. It gave me a sort of sad feeling ... I always felt that – not that it was haunted, but that *feeling* of – that it was left by someone under duress. It wasn't left by someone *happily* ... And that sometimes doesn't settle well. It's funny, I know you can do various house clearances – you know, you can make yourself feel better by clearing the house, burning incense and things like that. Which I do. Or a ritual when you go into a house. Which, when I go into my next house, I will definitely do that ... Just – cleansing the house.

But she had felt reassured when the previous owner had been spotted walking down a local street, after which she readily embraced a rather more reassuring narrative – of someone heroically surviving misfortune:

> She's obviously still *here*. She *survived* – you know. Someone spotted her. ... That's happier. It hasn't finished her off. That's a nice thing. When you think about it, people go through ups and downs. ... It's quite inspiring. Someone's taken a chance and started a business. 'Ok, it's failed, I can't afford the apartment any more. I'll do something else'. That's quite inspiring.

Jack in Bristol also turned to material rituals – involving cleaning and painting – in an attempt to exorcise the previous resident's presence:

> I mean I've repainted throughout ... I've cleaned so many times. I mean there was 40 years of filth, and the walls were just sort of, *grey*. ... It was thick with grease ... Cleaning and painting. What do people do? Like exorcising. I think they run around with little bells or something [laughs]. ... This is a sort of *milder* version, just painting it. ... And that's the first thing I did, was clear out this, vacuum. And I vacuumed and vacuumed and vacuumed. I don't want any of her *anything* there so I vacuumed everything.

He was also considering renting the house out to complete the process:

> I think it would be quite good for me to go away for a bit and just – because it was such an unpleasant experience, buying the house and moving in – that one of the reasons I'm thinking of renting is just to have another presence in and have some happiness in-between, you know, which completely disconnects me from the previous owner.

The rituals of cleaning and cleansing – ridding the new home of the recent resident's material and psychic presence – had not quite worked; there was a need for a more dramatic solution to replace and recharge the lingering atmosphere. As Jack's example suggests, this becomes particularly important in the aftermath of more negative experiences.

But reminders of previous inhabitants can be hard to shake off. Anna, for example, described how a plant inherited from the badly-behaved vendors kept on returning every year: 'Oh that's the Jones' plant come back to haunt us. ... That's the Jones' flower – flowers that were here that keep coming up [laughs] from when they – [laughter]. ... We can't get rid of them. We've tried to get this out. *Yes.* We've had this out – I'll show you the photos of when we had the garden redone. And absolutely *everything* was out. And this *still* comes up – after *26 years*'.

The plant reflected the vendors' rudeness – intruding into Anna's space. She was incredulous, but made light of it with a joke.

Elsewhere, there was the common issue of receiving previous residents' post – also seen as intrusive although offering the opportunity to eavesdrop into past lives. For Samantha in north London, it reflected the dubious character of their vendor, the architectural salvage man with his tall stories about the interesting objects left behind in the flat: 'We get slightly threatening letters for J. Unpaid bills. That's the only way we know that they're threatening, is that they are *red* bills ... bills he's not interested in paying'.

Karen created a profile of the last resident through the post which continued to arrive many years later:

> I can give you his name [laughs] ... And I think he had a girlfriend who lived here for a bit. And I think at some point they were also running a company called Z.M. – out of here. This is all based on things that come through our door. It's weird to know. You kind of pick up a *picture* of somebody by the kind of post – you know, like flyers for antiques fairs. So you kind of feel curious about them ... He remains *there* in a way ... I occasionally open it thinking – 'Oh, I might go to that antiques fair' [laughter]. You know, 'if you keep sending your post to my house, then I'm going to –' [laughter].

She might have finished her sentence by saying, 'take ownership of it'. The boundaries of ownership are blurred as the previous resident revealed fragmented aspects of himself to the new owner, in what was now *her* home. This elicits a range of feelings – Karen felt his 'presence', was curious about him, even joked about having interests in common. But it also felt 'weird', and she was annoyed at his thoughtlessness:

> He definitely didn't change his address properly. And I was returning to sender, you know, things from – Inland Revenue and – you know, the types of post that one would probably want to know that the person no

longer lives here [laughs]. You know, we've had one or two bailiff's letters, kind of serious, like – 'Maybe you should have told some people you've moved house' kind of letters ... It's just a failure to organise yourself ... I'm just a bit annoyed. I'm just *inconvenienced*.

The unwanted post also set up a moral obligation, requiring decisions about what constituted correct behaviour. Karen decided on a two year cut-off point, after which she no longer felt obliged to return to sender. But she still felt the need to justify this, arguing that there was no 'real personal post' which 'I have thought, "oh, he'll be really upset not to have received"'.

The post of others becomes a form of unwanted gift, a rather different experience than that described in Marcel Mauss' (1990 [1925]) work, where traditional rituals of gift exchange were considered part of developing 'social bonds forged through systems of reciprocity, obligation and kindness' (Blakely and Moles 2019). Despite the 'receipt' of a gift being considered 'itself a gift of acknowledgement', this is less necessary for *unwanted* gifts – post, plants, rubbish. In contrast, the blessings – the friendly notes left by vendors – *do* set up an obligation to reciprocate which can itself prove burdensome in a context where some participants wished to maintain boundaries. This is reflected in Julia's complicated response to her vendor, whose note she had felt the need to respond to. Having not done so had troubled her, leaving the gift incomplete for having not expressed her gratitude for it:

> One of the things I felt terrible about was that I never responded to her letter because I was so busy when I got the house – for about *16 years* [laughter] ... I felt so guilty and it had haunted me for ages and I couldn't find her letter and I was so annoyed ... I felt so bad about the fact that I hadn't responded because I thought it was such a kind and generous thing to do. ... When I *found* her letter – 'I'm going to do something now, right, before I go off the idea' ... I just said to her, 'I've really enjoyed living here. Thank you so much for selling me the house' ... I felt I'd def-initely done the right thing then – reaffirming the connection.

After so many years, Julia was taken aback to get an immediate response from the woman, who happened to be visiting London. This created a new obliga-tion – but she made an excuse for not offering to meet her or ask her to the house. She added:

> She said, 'By the way, you know, if you ever fancy a house swap, because that's why I'm here'. And she sent me the details of her home in America, which is a lovely place. And I said, 'That sounds great'. And of course it would be quite funny if she – It would be so *funny* right? ... I'd know she'd look after it ... That would be very *odd*. I think it would be very odd for *her* ... I'm not sure I want to do it anyway.

After 16 years the woman is suddenly – on *the same day* – in her life, near her doorstep, and offering a house swap. This suggestion of a further transaction – of moving back into her old home briefly as part of a holiday – appeared to re-entangle their lives despite Julia's brief, polite email fulfilling her need for completion. Although she joked that the woman would know her way around her house, she described how the idea 'would be very odd' – preferring to emphasise that it would be odd for the woman *herself* to return to her old home. In turn, she mused over how such a relationship needed to be kept at a distance – the 'fantasy' of the vendor being better than the reality: 'I mean obviously, when a person becomes *real* – I mean when you meet them, even if it's only by email, you do realise how little you know of them really. Everything you think about is a fantasy, right?'

Despite the importance of their initial exchange, it was preferable for Julia that the woman remained symbolic, not a real person. The generous initial note had helped to smooth the passage from one owner to the next; the blessing was about gifting the home. Receiving and acknowledging such a gesture, in turn, was about respecting that the woman had also felt a sense of belonging to the home. Returning for a house swap threatened to destroy the fragile potency of this ritual.

For other participants, it is not always possible (or desirable, as will be described) to maintain distance from the previous resident. For Jane and Nigel, the situation was particularly claustrophobic – the vendors, inheriting a larger house after a family death, had moved *next door*. Jane and Nigel were polite about the situation, describing them as 'very good neighbours' who 'come in and feed our cat and things like that'. But there was no hiding their awkwardness. In part this was because the house had been in a poor state when they moved in, requiring 'more work than anticipated. We've changed it quite a lot'; in particular, 'some of his work wasn't that brilliant'. They joked that they couldn't 'moan' about this to their neighbours: (Jane) 'We could go and say, "gosh those people who lived in here before!" You know ... I think it would be better not to have the people [next door]'. But in addition, the vendor often commented on work being done on the house and had taken a proprietorial interest in the renovations. Jane explained:

> It's very *strange* sometimes. Because you feel like you should almost ask permission to do something [laughs] ... But quite often if we *do* do something, the man – he'll say, 'Oh, *I* did that'. Or something. Or, 'I remember putting that window in' [laughs]. Every time we have workmen here, and obviously they go outside for a cup of tea and a fag [laughs]. And he will *say something* to them, like – but not like, 'What are you doing?' He'll say, 'Oh I remember putting the last bathroom in', and things like that. It's quite comical really ... It's a strange relationship next door.

The neighbour's continuing presence and watchful gaze also proved to be inhibiting:

NIGEL: You still feel constrained about making changes.

JANE: It's odd isn't it? Even though we own it. We pay for it.

NIGEL: We've got plants in the front garden we don't really like. But *they* planted them. And quite often when I go by, they say, 'Oh, that's looking good, isn't it?' So we think, 'Oh, we can't get rid of that'. We don't really like it, but we don't feel we can get rid of it … You do feel a little bit like that. The garden hasn't changed very much.

In contrast to Anna's attempt to dig up the flowers that kept reappearing, Jane and Nigel had gone beyond the point of being able to change the plants, having agreed too many times they were 'looking good'. There was no animosity here, just an acknowledgement that their politeness and lack of assertiveness had led to a commitment to constraint, allowing the neighbours too much power in repeatedly performing their prior relationship to the house.

Returning to previous homes

Jane and Nigel were aware of the neighbours' continuing sense of familiarity with their home, an example of a common aspect of people's broader sense of affective ties across a range of past and present homes. This experience influenced people's belief that previous residents might continue to feel a connection with *their* home, just as *they* do with past homes. Ben in south London, for example, regularly walked his dog past his previous home only half a mile away. It contained many family memories: 'I'm very emotionally connected … because it was where two of our kids were sort of *born*, I guess'. He felt annoyed by the changes that had been made – another focus on a front garden plant, still felt to be *his*. He simultaneously mocked his own attitude:

So when I walk past there – you know, I can remember what we were *doing* in the house – the kids running round and down the road outside. And, and I take quite a proprietorial look, thinking, 'Oh well', you know, 'he's not cutting my wisteria. He's not looking after the wisteria is he? Oh dear'. You know, 'the paint's peeling'. And, 'What's he doing *there*?'.

He later attempted a more measured response, but fell back on a judgement:

The wisteria – it was a very *fine* one. Anyway, people have to do what they have to do, I suppose. And if you run out of money I guess [pause] – the wisteria goes, or the outside of the house goes. Um, and it goes really very quickly. Very, very quickly … And also because I think we did quite a lot of work there. And – we opened everything out and made it look bigger and lighter. And now they've sort of put *curtains* on it, and everything looks kind of small and I'm thinking, 'Oh, I just want to take the curtains down and – open it up to the light' [laughs].

The curtains, like the plants on view, became a symbol of bad choices he wished to reverse, removing them in one swift, easy gesture in his imagination. Other participants also judged changes made to previous homes. Pam in Yorkshire, for example, described looking at photos of her previous home on an estate agent's website when it came onto the market: 'And they built a huge extension. And I was looking at the photos and tutting like mad because I felt, it felt: "Oh, oh, they've completely overdeveloped it, that's far too big". You know. I was judging what they've done very negatively and feeling upset for the house'. She also recalled contacting the people who owned a previous home. She wanted them to know that she and her partner had 'put the fireplace in … I was saying: "I put that fireplace in" [laughter]'. She acknowledged but self-consciously mocked her pride in laying claim to work she had carried out many years previously.

Reflecting on his previous home, which he also regularly walked past, Chris noted he was 'part of the history', but that he had no right to interfere. He played down his imagined reaction to discovering the old windows replaced by *plastic* ones, an emotional response which, however, quickly found its limit:

CHRIS: They've still got the original small pane sashes. And I'm quite gratified. But I mean I've got no – if they wanted to put in plastic windows …
I'd be slightly irritated every time I went past but it's nothing to do with me anymore.
JOSIE: You have to let go.
CHRIS: But it's *hard* [laughs].

Adam in east London and his wife Louise described pretending to be interested in buying their old flat when it came on the market in order to take another look, 'just really being nosy'. The flat was meaningful to them because their daughter was born there: 'It was a very happy time, I suppose, for us. That was the first property we bought, renovating it from – we never had any money – it was all done with second-hand stuff and it was "shabby chic" as they call it now [laughs]'.

Louise recounted a vivid dream before the visit: 'What was really strange – I had a dream about going in the place. And in this dream there was a red bedroom which was at the back. And when I went in there, it was a red bedroom. That's the bit they sort of built after we left, didn't they? And extended it out the back. Very strange dream'. Although they noted what had remained – the stove they had put in, the cat flap – the dream focused on what was new: the extension with the red-coloured room. It reinforced not nostalgic memories of their time in the house but what had changed – an emotional dissonance which was as 'strange' as the fact that the dream was a premonition of how the house had evolved after they no longer belonged in it.

A continuing feeling of connection to past homes led some participants to avoid returning to them. Susan sent her daughter to visit her old home when invited, feeling too protective of her memories to go herself:

My daughter went and she said, 'Oh their baby's in my bedroom and it's really nice'. And I couldn't do it. Because it still feels like it's *my* house … It's because I have this attachment to it. I don't want to see someone else living in it and the changes that they've made. I don't mind someone living in it. I don't want to see the changes so that in my mind's eye it still looks the same as when I lived in it.

Gemma felt the same: 'I was always told by an old lady, "Never go back to where you've lived in" – because they *wreck it*. You imagine it as you had it. And someone else will move in and – I did have a chance to have a look but … I lied and said I didn't have time. Because I remember her saying, "Never go back. Never look back". Because it would no longer be yours. It would be in your memory. It would be somehow *spoiled*'.

For some, a return to a childhood home can also be a painful experience, as Ben discovered after returning to the house where his mother had died when he was a teenager. He recalled: 'I don't know if it's a good thing or not really … It can be very odd. It depends on what your emotions are. It can be quite disconcerting I think … It was just a very strange feeling'.

Previous residents visit

Participants thus understood that emotional relationships with a previous home do not end when people move on to the next home. This in turn had an impact on responses to *previous* residents who visit participants' current homes. As with the intense process of exchanging one home for another, so too are such brief encounters with past residents dictated by social mores, required to negotiate such assumed overlapping feelings of belonging. These visits elicit a range of feelings. Cathy, as described, found it 'spooky' and 'surreal' when the family of the famous photographer visited. Jacob in the southern English town had a particularly personal reason not to invite back a previous resident – reflecting a relatively common form of 'small town' coincidence, his ex-wife and her second husband had owned the same house previously.

But just as some participants avoided returning to a past home, they themselves could be met by past residents' own reluctance to engage with *them*. Yolanda in the north London suburb attempted to contact the daughter of the original owners to seek photographs. She wanted to get a glimpse of them and of the house's original décor. The daughter had responded politely at first, but didn't follow this up. Yolanda felt she couldn't chase her any more:

She did write back and said she needed to find the photography. And then I wrote back – I think it was a letter. And she said: 'I need to find them', and she never came back to me. And I thought I can't write again, you know. I can't just keep writing … I have considered it would be impolite … Maybe she thought I was just being nosy or something. And I did say to her, I am a design historian and that's why I'm very interested in what my own house – yeah. But, you know. 'I wouldn't let anybody else see them'. All those reassurances you can give. And it was only for my own personal interest. I would just love to see them.

She wrote twice but couldn't pursue it further, aware she might be seen to be intruding but defensive that her 'legitimate' historian's interest in her house had failed to persuade the woman of her serious intent. The importance of correct etiquette in communicating with the past resident was intensified by an assumed need for the woman to safeguard childhood memories and personal mementos of her deceased parents. The situation reinforced the fact that, however connected these separate lives were through the sharing of the same home over time, they remained – or sometimes needed to remain – strangers.

For others, however, the relationship with more recent residents can be more open and celebratory, particularly if, as in Lillian's case, the new resident is enthusiastic about the design efforts of the previous one:

It was very, very eccentric and wonderful and I loved that house. That house feels more like my home than any other home. It was absolutely beautifully done ... with great love. And the guy who was selling it had to sell it because of divorce and he'd done all the work himself. He was a West Indian guy, and I went in through the front door, just went in through the front door and I said, 'I'll take it'. I said, 'I love this house'. I loved all the features. I loved everything in it. And I was admiring it and I said, 'I love this house. I'm going to look after this house for you. And I'm not going to change it'. And I said, 'If I do it will be sympathetically done and I can assure you that the love that you've put in this house I will also keep'.

This emphatic iteration of appreciation for the love she felt the man had 'put into' the house became an affect she wanted to 'look after' and 'keep'. This was less about respectful etiquette in dealing with past residents, or the need to reject past choices, or to feel inhibited in doing so. Her enthusiasm for the design aesthetic meant she was happy to inherit the home intact, and to be a custodian of it for him – to 'look after this house *for you*'. Before she moved from the property, she invited him to see how she had 'enhanced' it without having 'taken anything away from it' as she had promised: 'He was thrilled. And the people that came and bought it from us ... wanted to keep it basically the same as well'. Here there was an unusually seamless passing on of a material legacy through a shared affect bonding previous and new owners; there was no need to stake new claims or stamp fresh marks on the house in the act of making home. Her invitation to the past owner to visit was a final act of respect, acknowledging his contribution to it and proving she had kept her word – her enhancements had not fundamentally changed the 'love' he had put into it.

Seeking approval from previous residents

Elsewhere, previous residents are often afforded particular respect for having lived in a home, and there can be a kind of empathetic anxiety when imagining how it might feel to view the changes through *their* eyes. In some cases this manifested in relation to a desire for approval. Pam in Yorkshire, for example, described her pleasure in gaining approval for what she had done to the house from those that were 'born here' who 'have a legitimacy that I feel I lack'. She explained that once or twice a year parties of people who used to live and work in the dale come back to visit:

> There are a couple of brothers and their mother who I have met. ... The family kept chickens here when it was a ruin, but they said they really like what I have done to the house. That gave me great satisfaction. ... The root of that is the fact that if they were born here, they have a legitimacy that I feel I lack. And that they approve, that they like what I've done, means I can borrow some of their legitimacy. And it also ... [means] that I've created something that the people with legitimacy feel is appropriate.

Susan also described feeling reassured and delighted when a previous resident came to visit and appeared to approve of the modifications she had made. She had been introduced to the elderly widow who still lived locally and invited her to look at the house:

> She's seen the changes we've made, which I was interested in, yeah, because she saw it before and after ... I knew she must have lived here at least 20 years ago. ... And she's very, very old but very compos mentis and not interested in houses at all [laughter]. But she came round and ... she said, 'Oh, it's so much better!' [Laughs].

She described her pride in being given approval for changing the kitchen, the fireplace, and (more controversially – and therefore more importantly) for felling a tree in the garden:

> Yes, she said, 'Because the kitchen used to be in that corner' – there's a little room in there and it's tiny. She said, 'I don't know why we never did that', moved the kitchen out. And there was also a massive sycamore in the garden, which we took out. And she said: 'Oh, and you got the sycamore out', which we had a lot of trouble to get out. She said, 'I'm really sorry, I do feel bad about that because that was there when we were here and I should have taken it out, because I knew how much trouble it would cause' [Susan had in fact got into trouble with the freeholder Estate for removing the tree, so the woman's apology was particularly gratifying] ... She thought this was great. And we took an old fireplace out in the living

room and replaced it with one that seemed to us to be more fitting. And she said: 'Oh, that's a much nicer fireplace than the one we had'. I don't think she was being polite. I think she genuinely thought it because she's not that kind of woman. You know, she's quite a forthright woman … [It was nice] to give me that affirmation. Yes it was. That was really nice. I felt like that was – *wow*. I was *really pleased* that she said that.

Older residents visit

Other participants also described meeting older residents, either those who turned up unexpectedly or were put in touch by neighbours or local acquaintances. These encounters are often described as benignly interesting conduits of oral history, offering insights into the previous layout of the house and local changes, just as the 'how to' house history books suggest (Brooks 2007; Bushell 1989). Jane and Nigel, for example, recalled an 'elderly gentleman' who 'knocked on the door' and said, 'Do you mind if I come in and have a look round? I used to live here' … He started to tell us, "Oh that's where *that* was. That's where *that* was. *That's* gone". And all that sort of stuff … He explained about the coal cellar. So it's interesting to get some – first-hand account of how it has changed'.

Some participants expressed excitement in hearing the stories of first-hand witnesses to historic events. Chris and Josie, for example, were put in touch with an elderly man who had lived in their house during the turn of the twentieth century. Josie had not believed her colleague at first when she was told about him:

I said, 'Oh I don't think so because [previous resident] Miss H was here from 1926' – thinking there probably can't be anybody still alive who'd lived in our house. But actually there was … It was Mr W … So I phoned him and said, 'Would you like to come around and see your old house?'.

The chance to be hospitable to a previous resident – whose relationship to the house predated theirs – was combined with a desire to gain insights and testimony ('I mean partly because we thought he'd be interested but also of course we wanted to hear his stories [and] I wanted to give him the opportunity to see where he'd lived as a child'):

He was here from 1916 to 26 – in quite sad circumstances. Because his parents had lived at … [a] farm up the road, which is now a hotel. And in 1916 or 15 – I've got the details somewhere – his father was killed in a farming accident, leaving his mother with eight children. She had to down-size and she rented this house for ten years, let a suite of rooms to the curate from the church – and so he was here from the age of four to 14, which were quite formative years for him. And I mean he was just lovely, wasn't he? … He remembered the end of the First World War. He remembered some Australian soldiers hanging out of the windows, the end house in St M. Street, shouting, 'The war's over'. This is the *First World War*!

CHRIS: I just thought, you know, what a connection with *history*! There's somebody telling you a first-hand account of something that happened nearly 100 years ago. And – it's a thing about putting sort of timescales into perspective. Somebody's memory goes back that far.

The elderly gentleman was a figure from 'history', representing an uncanny intermingling of times long gone and *now* – the stretch of memory recalling an intimate local moment reflecting national intensity (Kean 2010). At the smaller scale of the interior Mr W drew a plan of the house and garden as he remembered it, describing what appeared to him to be more similarities than differences, the embodied encounter with the distance of history matched by the domestic repetition:

> In the dining room I had the old slat back chair in front of the range, and he said, 'Well', he said, 'my mother had her chair *just there*'. So there's, you know, that was quite nice really [laughs] … He walked in and he couldn't believe we'd got the chair right *in the same place* … He actually was quite surprised how little had changed.

Such visits were generally uncomplicated forms of hospitality, mutual openness and respectful interest. The distance of time – even when it *felt* so close – carried none of the fraught manoeuvres of negotiated belonging apparent in encounters with more recent residents. Nonetheless it is still bound by unwritten rules of etiquette. As Chris reflected about Mr W's enjoyable visit, he would not be comfortable with *too many* visits: 'Not repeated visits. You do it once or twice out of curiosity', he stated. There was still the need for distance.

This became an issue for another couple, Harvey and Rochelle, in their Georgian house in central London. Whereas the degree of the span of time (eliciting excitement about the mixture of differences and similarities with *now*) meant that Chris and Josie felt no self-consciousness, or need for approval, from Mr W as he viewed his childhood home, the timbre of Harvey and Rochelle's contrasting experience suggests that the changing socio-economic context can have an impact when encountering people from the more *recent* past. Their neighbourhood had only started to be gentrified, a mixture of social and private housing, with many of the older houses being turned into small hotels. But still, the contrast with the past of a few decades ago was stark. The area, known for its brothels, had been very poor, and one day someone from those times knocked on their door. Harvey told the story:

> Probably about four years ago our doorbell rings one day. And there is a guy standing on the doorstep. And he says, 'I'm sorry to trouble you

but I used to *live* in this house'. And I said, 'That's interesting. Come on in'. You know. And to cut a long story short, he is now a retired railway engine driver up in Lancashire.

The man's parents, he explained, had run a brothel a few doors down the street, during the 1950s: 'They couldn't keep him in the brothel and he got farmed out to an aunt in Ireland. That's where he grew up. But he used to come back *here* all his school holidays, because his grandmother was living in *this very room*'. When he passes through the area he 'bangs on the door and says hello – which can be at quite *odd* moments you know, but um – he's been back several times'. The hint was that the man was at times a nuisance, breaching that social etiquette about repeat visits. Rochelle described the man's 'rough' sounding voice as someone who no longer fitted into the area:

[Semi-whisper] It was quite funny. We had some builders working here at the time. And they heard … the doorbell ring and they heard Harvey talking to a fairly *rough* sort of chap [laughs]. They asked me whether Harvey was alright [laughs]. They were quite kind of *protective* [laughs].

HARVEY: Well old N. He's kind of an odd guy. But he's very *decent* you know … I mean, you know, when he rings the doorbell it can be a convenient or an inconvenient moment, you know –

ROCHELLE: [Laughs] last time it was *inconvenient*.

HARVEY: I just said, 'It's nice to see you, but I'm afraid', you know, 'we're busy' – and so on.

ROCHELLE: 'We've got visitors'.

Despite describing strategies for dealing with the man's unexpected visits, Harvey said he had been 'extremely interesting' and 'useful'. They had taken a long walk around the neighbourhood and he had pointed out 'quite a few things that had changed and what it used to be like' and answered questions about the house. Rochelle also became thoughtful: 'I think it's a bit mixed for *him* actually. Because it was very poor when he was living here':

HARVEY: I think he had quite an emotional reaction to it all. Because it was all his childhood memories. A lot had changed.

ROCHELLE: He brought his son didn't he? I think he wanted to show his son … He couldn't quite *let go* … I think I felt [laughs] slightly *awkward* about it.

Their narrative weaved between factual accounts of his visits, how the area had changed, and their attitude towards *him* – their having to deal with the occasional intrusions into their private world by someone who 'couldn't quite let go'. Despite their sympathetic understanding of the man's need to visit, and interest in his story, Rochelle admitted to an awkwardness about the socio-economic differences he represented, the consciousness of their status reflected in their comments on his manner and their self-consciousness about *his* awareness of what had changed – their immaculate home, restored to its elegant original era, a symbol of the area's gentrification.

Passing on practical information

As these examples show, people's continuing emotional ties with past homes (and what they represented for the narrative of their own lives), is matched by a consciousness of the continuing emotional ties of others to *their* homes. Harvey and Rochelle's brief encounter with a past resident reflected one end of the spectrum of engaging with a home's past within living memory, Chris and Josie the other end. Both involved past residents as living people with their own stories and emotional responses, expressing the impact of different circumstances and the way some things change and others stay the same. These encounters became part of extended forms of sociality involving different degrees of negotiation of ownership and belonging with others. They also suggest extended *temporal* impacts on home, including, elsewhere, people's *future* relationships with their homes. These were either projected through imagined scenarios or enacted through particular practices, often at the liminal point between staying and leaving. For some, the thought or experience of leaving reinforced the temporary nature of belonging, but at other times reflected their own difficulties with 'moving on' and strategies for maintaining contact and degrees of control.

Within the process of passing a home on to the next resident, participants also revealed their social values and ethical beliefs. For some, this involved a *duty* to pass on, or to offer information about the house. Lillian, on the main road in south London, was planning to carry out house history research, and envisioned a formal passing-on ritual: 'whoever buys the house after us, it'd be nice to give them a presentation'. It was also important to maintain the memory of the house through assembling its history into an accessible object. As one of few older properties remaining in the street, she was anxious that eventually it would be demolished in the local frenzy for new developments:

> If for example they decide to demolish it – I can see honestly that they might buy this one, the next one and the flats and make an enormous development, a skyscraper on this place. I haven't got control over it. But I do have a history. So that people in years to come can look back and say, 'Well, that was there'.

Others also planned to leave information on the house's history, but were aware that future residents may *not* be interested. Leah said: 'I'll make this [folder] tidier and send them an email with it on a word document. ... I don't know if they're interested ... if they're not, they're not. They probably will be'. Anna wanted to write a book to state 'this is what's happened in this house – through time. And this is the story about this fireplace'. She would leave it behind, even if the new owners decided to 'throw it out', but assumed that they would also want to know the small details – the facts about who did what, when: '*I* wanted to know – who *was* it who painted all this coving brown gloss? And what was the story behind *that*?'

This passing on of information would be a different form of 'gift' to the next residents. Participants assumed that it would be interesting and aid the process of making home, but they also acknowledged that this might not be the case – and that other people might not share their interests or concerns. Joy, renting the east London flat, was more protective of the scrapbook she had created, partly because it contained her own drawings of the communal garden – including of an old tree she had loved that the landlord had felled. She said: 'I don't just want to leave it and be chucked in the dustbin. If I knew the people that were coming in here was interested, I'd leave it. Otherwise, it comes with me [laughs]'. Here there is a *condition* that the next residents needed to be interested and appreciative or she would not offer it.

For others, it is the practical knowledge rather than the house's history that needs to be passed on, particularly the more hidden elements. The withholding of such knowledge is considered unethical. Jack in Bristol argued that understanding how a house functioned was part of the process of becoming acquainted with it, a ritual of 'just going through everything, going into the attic and all sorts, you know'. But his vendor, who had been unpleasant towards him, had chosen to *not* pass on information: 'I've got no details. That's what's really upsetting … She showed me *nothing*. She told me *nothing*. … I don't know how to work the central heating. I didn't know where the stopcock is for the water. I know no history here'.

Harvey in central London, a retired engineer, had recorded the hidden technical workings of the house: 'If we do some work on the house, I'll take photographs. Because if you ever have to say, "My god, where did that pipe run? What's underneath the floor when we did that?" Or, "How did we do that?" My main motivation was to have a record of what was done because if a problem occurs … and you can't remember it, photographs might – *just* might help. Such an easy thing to do. I'll pass them on'.

Knowledge of a home's more quirky elements required hands-on demonstration. Chris and Josie pointed to the 'awful lot of quirks and foibles' in their house, including the fact that 'It's almost impossible to open the back door' (Josie). Leah and Alex also described the quirks of the house that Alex, a builder, had renovated over decades. His exclusive knowledge of how the house worked became a focus for the enactment of his continuing feelings of belonging and identification with it. The couple, as described, were in the process of selling the house. They had chosen an 'ex-pat' family to buy it and the husband was renting the self-contained flat at the top of the house whilst organising for his family to return to England from the Middle East. Alex planned to maintain his presence in the house by being on hand for 'all his DIY SOSs':

He's got to learn all the oddities and … where everything is. A big house has a lot of bits and pieces in it. … He couldn't turn the shower light off [laughs]. The shower room light – you don't give it a gentle pull. You have to – if you pull it gently, it dims. So [laughs] he hadn't sussed it out. He asked me: 'How do you turn this light off'. I just went [clicks tongue], like that … I've got lots of knowledge, you know. Silly things that, I don't know, *might* go wrong. It's the drains that *might* block. And the leaves which *might* get behind the garage gates if you don't. … And I'm quite happy to do that for him. … We've become friends anyway, which is, you know – *nice*.

Maintaining contact

As Alex hinted, befriending the new resident is a useful way of maintaining contact with a previous home. Alex also wanted someone who appreciated his renovation work and was thus less likely to want to change it:

> The guy buying it *loves* all the Victorian features – absolutely *adores* all this woodwork and shutters and – you know. He loves all that stuff. Coving. You know … I spent years and years of my life doing – you know, improving the house. … And for somebody else to take it to the next level, to make it a nice family home – is important to me. Yeah. Yeah.

Even better, he added, that the buyer had promised to seek *approval* in advance for any changes he might want to make: 'Well I'm going to get call backs from this guy [laughs]. You know, he wants me to come round and approve the stuff he's doing'. This level of promised accommodation to Alex's wishes appeared to have been cultivated through their personal friendship. As ex-wife Leah mockingly described: 'They've been drinking wine out in the back garden together [laughs] – having a *romance*'.

Others also attempted to maintain contact with their homes through selling to people they could befriend. Rita admitted to this when forced to sell her inheritance, a house which had been in the family since 1912: 'I still love the house and I feel, you know, as though it's really much more mine than the current owners'. The original owner, her great uncle, worked as an artist from a studio in the garden. Rita sold the house to artists who had promised to renovate the studio – 'to use it in the way it had been used'. This had felt 'very appropriate – *right*'. She had made an effort to keep in touch, even staying in the house with them, and exchanging one of her great uncle's paintings for one of theirs: 'We did a swap. It was just a nice thing. It sort of consolidated our bond. … I made sure the relationship continued'.

But eventually the artists had moved on and Rita's attempt to befriend the new owners had been less successful – she had lost control over who lived in the house and therefore her contact with it:

When I visited the village, I stuffed something through their letterbox, something about my great uncle ... the brochure of the art exhibition [of his work] I organised. And I said, you know, 'You might be interested to see this. And I will also be selling some of his paintings ... over the next few months'. You know, thinking they might be interested. ... They just wrote me a nice letter saying, you know, 'Sorry, we weren't here when you called. Hope to meet you at some point'.

They hadn't been in touch since, her frustration about their lack of interest not quite matched by an understanding that 'different people have different responses. But I mean, if I had just bought a house and people said, "Look, these are the memories, of the place, going back", *I'd* be fascinated'.

Leaving home to 'like-minded' people

For some the focus isn't as much on maintaining contact but a desire to sell to people who shared similar values and (as Alex suggested) a vision for how a house should continue to look. Lillian, who had promised the previous owner she would look after his home and only make 'sympathetic' changes, explained: 'Anybody that would buy this house would be interested in history ... I wouldn't sell it to somebody that wasn't interested'.

But anxiety about losing control is also acknowledged. For Pam in Yorkshire, it would be 'absolutely essential' that the new owners shared her ideas about what was appropriate for the house's age and aesthetic. She said:

I was having a conversation with [a friend] – what would I do if somebody came in and started musing about where to put the fitted kitchen? And I said, jokingly: 'I'd just tell them to leave, that I wasn't going to sell the house to them if they wanted to put a fitted kitchen in' ... If they come in and want to start changing the *personality* of this house, I wouldn't sell it to them ... Not that I'd *know*, of course.

Brian and Megan told a tale of how they were in competition to buy their house with a rival bidder who had offered *more* money than them but made the mistake of telling the vendor what he would change:

They'd had a nice brick pier built which the gate fitted onto. And he said, 'Oh, I'm going to knock *that* down because it's not big enough for my car to come into'. He'd really done everything wrong! [Laughter]. ... They were upset by this man. And when he offered them more money than we were able to pay ... they felt instinctively, they didn't want to sell to this man ... So that's how we came to be here.

(Brian)

Megan did not feel the same anxiety about future changes as 'it would not be my house any longer. So whatever anybody did, you just *shut off*. You have to'. But she modified this, adding they would need to be the right *kind* of people for the *area*: 'I would have to like them. I would have to feel that they were going to fit in – this *neighbourhood* ... being the *right kind* of people. If I thought they were real *yobbos* and *horrible* [laughs], I wouldn't want to sell it to them'.

Here it is the people themselves that need to be 'in keeping', to keep up appearances in the neighbourhood. The gravitating to those perceived to be similar in behaviour, values or interests is also reflected in decisions to buy. Finding connections created a reassuring familiar sense of homeliness – for some a key way to smooth the transition from one home to another. In two examples, this focused on a shared interest in books. Abigail in the Georgian flat in north London felt an immediate connection between herself and the vendor, noting that he shared her interest in esoteric beliefs:

> And then I see this man who I don't know, I've never set eyes on, who shakes my hand. We immediately like one another ... I remember saying to him, 'What a fascinating collection of books'. He said, 'Yes, I'm very interested in comparative religion and philosophy and Eastern philosophy'. And I said words to the effect of, 'Yes, so am I' ... That was the clincher, if you like ... I was going to live here.

Samantha and Jake in their north London flat, full of salvaged objects, also described feeling a 'sympathetic atmosphere' because of the knowledge that the previous tenant had loved books, helping to aid the delicate exchange of senses of home from one stranger to the next. The vendor told the story of the man, a refugee from Eastern Europe who had rented the flat from the 1940s and became a well-regarded specialist bookseller. Samantha said:

> I found out a fair amount about its occupant ... some of which encouraged us to buy the flat ... I believe this room was full of books. [We have] an inordinate amount of books. I like the idea that he was interested in books ... this flat was *full* of books ... there was a feeling that this was [a] place that – will have a – *sympathetic* atmosphere.

Cultural differences: those 'not like us'

Jake also described other 'sort of coincidences' since moving into the flat in relation to 'little connections' discovered with people in the area. He considered this a 'certain kind of narrowing' which reflected synergies between lives: similar people were attracted to a place like magnets or 'cubelets' (referring to 'robot' cubes magnetically stuck together); this form of 'coincidence' accentuated similarities between people. But there were hints at social engineering – of people choosing or *being chosen for* a home because they already had the 'right' profile. Or as Megan suggested, *not* choosing others who aren't seen to *fit* as well. Thus the act of moving in or on from homes, and the various aftermaths of such a process, involved a variety of responses to encounters with people who were considered similar to oneself, or not similar *enough*. This weighing up of how strangers should be judged in the course of such brief, but emotionally loaded, encounters revealed assumptions, beliefs and potentially *prejudices* about difference.

Some participants, for example, fell back on national stereotypes to explain relationships to home history. North American-born Laura laughed at the 'British' desire to live in old, draughty homes with no *en suite* bathrooms or air conditioning. Americans, she said, preferred their modern comforts and were happy to 'just visit' old places. In contrast South African-born Jack in Bristol argued the opposite. British people had no respect for the material inheritance of their homes, forever replacing poor materials:

> Britain is chipboard ... You just stick up a modern chipboard thing and then two years later you tear it down and ... put in another brand new ... In Britain people modernise all the time and ... everything that's original would have gone.

In South Africa, in contrast, people 'prize' the quality of old things. Australian-born Anna in east London also compared the homogenous Victorian terraced houses in her street with her memories of Sydney, where the preference was for creating unique 'off plan' homes. These critiques embodied expressions of nostalgia for original homelands, although they tended to reflect *ideal* home types and experiences rather than acknowledge the range of circumstances in which people made home in these places.

Some participants also revealed assumptions in comments about other cultures, particularly those not originating from Westernised countries. William complained about the previous resident, who, he said, was an older Asian man with a much younger wife who had broken the estate's rules by squeezing three children into the tiny studio flat. He believed they had been 'happy to live altogether like that' because 'they'd been used to it – their *culture*, where they came from ... it was no problem for them'. Elsewhere, Karen described the socio-economic changes in her street, pointing to the remaining council houses as evidence that 'nobody used to want to live here'. She described an Afro-Caribbean family who had lived in the street for many decades, reducing the family to a signifier of the street's previous 'shoddy' reputation: 'I happen to know that 30-some years ago, Afro-Caribbean couples didn't easily get mortgages in nice neighbourhoods. Um, which ... sounds a bit awful, but that was true back in those days. So it must have been a pretty shoddy neighbourhood to live in ... in classy neighbourhoods, people were a bit prejudiced and racist'.

In turn, Alex's enthusiasm for his chosen buyer who had become friendly with him was matched by a *rejection* of another offer for his house by an Indian family. He recalled:

I thought it was going to be – was a *mass family* – thing coming in. You know, like brothers and sisters and – you know, a whole – it was an *Indian* family. An Indian family who, you know, obviously brought their brothers – and stuff around the house. You know, the guy who was buying it … he brought his brothers and his sisters and – I got the feeling they were just going to move *en masse* into the house. And it would have been a bit – [voice faded].

Both myself and Leah reminded him of the importance placed on returning the house to its original use, described in the census records as home to large families. Wouldn't the Indian family have fitted this criteria perfectly? Alex responded by saying that, for him, it was more that he had an 'instinctive' liking for the man he had chosen to buy the house, who had promised to preserve the original aesthetic he had painstakingly restored: 'He *admired* the house as soon as he walked in … He has the same taste as me, basically'. There might be an irony in the fact that the chosen buyer was an ex-pat returning from abroad, given the wider colonial context of his story. But if there is a suggestion that his choice was influenced by a cultural assumption – that the Indian family was unlikely to appreciate the house's Victorian aesthetic *as much* – the fact that he had found a buyer so easily fitting his criteria meant he did not have to make the effort to understand the response of strangers who did not, on the surface, appear to instinctively fit as well.

Challenging difference: sharing histories, taking care

What remains unspoken or unnamed can be a strategy of evasion, and accentuating some aspects of character or reasons for decisions over others might be a way to police expressions of feelings, beliefs or assumptions about others which might be considered controversial or awkward, as in the case of Miss H in Chris and Josie's home. This process is unlikely to be always a deliberate or conscious one. More complicated were encounters with previous residents whose paths overlapped more intimately, as happened for Martin and Carol in the east London square. They recounted a story about a man, Otis, who lived in the upper flat. He had since died and they had bought the flat, but for a time he had been their neighbour. They came to reflect that concerns about Otis might have led the previous owner to sell; what was and was *not* spoken or communicated to them became the focus of judgement. Martin explained: 'When we arrived here he was introduced to us. He was a black man, probably in his mid-50s. And the couple who lived here with three boys – introduced him a little – *coyly*. I thought nothing of it. He was introduced as a homosexual'.

Being told of the man's sexuality might have struck them as telling; they made a point of emphasising that this didn't concern them. But it was what the vendors *didn't* tell them that became the focus of the story – the fact that they had considered Otis's behaviour to be 'odd' – a different, more curse-like inheritance, the passing on of an unnamed problem:

> There were a couple of instances to do with post and things. And the estate agent said: 'They don't want any communication' … I got the distinct impression that one of the key reasons that they had moved was because of their concerns around him, which I felt was a bit *disingenuous* of them because, bearing in mind, we were moving in with a small child. Not that Otis was ever going to harm her, but nevertheless.
>
> (Carol)

Otis, they discovered, was suffering from early-onset dementia, triggered, they believed, by an emotional breakdown following the death of his boyfriend. Any concern about protecting children soon turned to concern for Otis. The couple attempted to understand and support him as his illness progressed, becoming fascinated by his life and speculating about the difficulties he had faced:

MARTIN: He was a very dapper guy. He'd go out with this umbrella – and his hat. Beautifully spoken. I found out that he was the lover of Lord F… an eminent reviewer of fine art … The feeling I get is that F's family said … 'What are you doing with a black lover?' Despite being homosexual, but having a *black* lover – and the family, you know, anyway wasn't very accommodating … Somehow I knew he came from Trinidad. I suspect his family just cut him off basically.

CAROL: But he must have been highly educated. He had had a job in the city, hadn't he? … He was a really lovely man.

MARTIN: But he was never quite *with* us. And as time went on, he got worse … I tried very hard to understand – about him and what problems he had.

For three weeks he didn't appear. The couple rang hospitals and got the police to enter the flat. Their anxiety soon turned to anger:

MARTIN: Then one day … I heard someone outside our door. And there was a guy with a beard and dungarees with Otis. Otis looking really – ramshackle, withdrawn … and nervous as hell. And I said, 'Otis, how are you?' He said, 'I'm all right. I'm very pleased to be back, very pleased to be back'. I said, 'Great. Great'. I looked at this guy and said, 'Well, who are you?' He said, 'I'm a social worker'. 'What do you mean, social worker?' He said, 'Well – we took Otis into care' … I said, 'Bloody what? You know. You didn't think to *tell us*? You didn't think to tell the neighbours?' … 'Oh, sorry about that'. 'Sorry? I've been worried'. Absolutely *mad*. They just *kidnapped* him. They just took him away … Don't die lonely, I'll tell you.

CAROL: He obviously just thought we didn't care … There was nobody to think for him.

The local council sectioned him and took possession of the flat, assuming, they pointed out, that relationships of obligation, care and support did not exist beyond familial ones. Chris and Josie visited him in a care home, complaining on his behalf to staff that he was 'the only black, the only guy, the only homosexual in a home full of old women ... I think what happened to him was unforgivable' (Chris).

By the time they bought Otis's flat following his death, the council had gutted it and thrown away his personal possessions. But Martin had already broken into the flat to find some items – books, hats and photographs – to take to Otis when visiting him before his death. They had chosen to 'inherit' some of these objects, keeping them as a respectful memorial to the man, the last remains of his life. Martin brought down from a shelf some books on art, written by Otis' partner: 'Effectively I stole some of the books. I thought, 'They'll just be chucked, or they'll end up under rubbish' ... You see, there's a personal note in there ... I mean that's – *silly sentimentality*. But, at the same time, they were part of Otis'.

The act of salvaging personal items was one of honouring the memory of Otis, a man who died in isolation, abandoned by his family and avoided by the other neighbours. In taking care of him, and his memory, they attempted to understand the man and his life – a form of 'domestic genealogy' not about protecting territory and enhancing senses of belonging but enriched by the histories of others in the act of taking care. This form of connection is not about seeking out the comfort of similarities to self but the capacity to accept the challenge that sharing home space over time with others might support forms of hospitality and inclusiveness.

Acknowledging that space is shared also offered possibilities for bringing people together. For Penny in the southern England town, whose house was on the main street, engaging with the home's past isn't just an exercise in imagination or fascination but creates an imperative to act in the world:

> If the thinking leads onto feeling, and leads onto action – yes there is value ... It leads onto a desire to know more and to understand more and to – *converse* more, to *relate* more both to the house and *other people* ...
> I really want to gather together in this house as many of the people who we know who have lived in the house, and the neighbours – as possible. In a 'friends and neighbours' sort of way ... To give the other people the chance to see the house, and to see the other people who have lived in it, and to – to talk about the house and their lives and anything else they want to talk about. The [house] is a sort of forum. It's a *meeting place*. It's a meeting place at all sorts of levels. It is literally, physically, a meeting place. But it is also a meeting place of things, *feelings*.

The value of thinking about the history of home for Penny was that it could lead to a form of *sociability* or relatedness which is not about negotiating private ownership but a vision of a *collective* and open exchange of experience. Owning

Figure 5.1 This collection of small found objects was placed into a plastic tub over time.

a large house in a prominent central location – one which had previously had shops on the ground and basement floors and where many people still living locally had either lived or worked – suggested such an opportunity. But there was no agenda or particular desired outcome behind creating a forum for sharing memories, stories and anecdotes – it is merely to 'see what would *happen*':

CL: You want to gain information – to know the house better?
PENNY: Not, not in a *greedy* sort of way … Just to see what would *happen*. Just to see what it was like.

Conclusion

The more immediate and sometimes complicated – if brief – encounters with recent inhabitants required different forms of negotiation in order to complete the 'transfer' of home from one stranger to the next. The exchange process is often key to the ritual of emotional as well as legal transference, but continuing relationships – with both past and future residents as well as previous homes – suggest that such a transfer is rarely complete. This is dramatised by the continuing unwanted 'presences' of past residents in the form of what they leave behind.

Leaving past homes and entering new ones is also a negotiation of different forms of *power*. Past residents can confer blessings – positive gestures, practical advice, expressions of care – or place curses – the withdrawal of information and other ways in which an already-fraught process can be made more difficult. Buying and selling to 'like-minded' strangers and enhancing feelings of connection between those coming immediately before or after can smooth the process of creating or maintaining feelings of being 'at home'. This is a strategy for mediating the exchange process, and a useful, if limited, way of ensuring one's own enduring presence when preparing to leave. However, in the same way that the assumed differences and similarities with earlier residents need to be managed as part of self/other relationships involved in experiencing different 'strangers' in the home, this is also reflected in the complexity of encounters with more recent ones. As shown, there are times when the desire to seek connections with others by accentuating similarities can also lead to the rejection of others and reveal or reinforce degrees of unacknowledged prejudice, assumptions about other lives or anxieties about difference (Ahmed 2000; Crang and Tolia-Kelly 2010).

There is perhaps a greater need to police the boundaries of encounters with more recent residents in order to maintain senses of belonging – keeping others at bay, *particularly* past residents who appeared to have continuing emotional claims on a home, just as many participants had themselves. But there are other examples showing that the ongoing process of establishing a sense of belonging to home can be less defensive, where home is seen as a space for more welcoming and inclusive encounters, for bringing people together to explore different experiences, or for engaged forms of custodianship involving understanding and care for others – such as Martin and Carol's 'inheritance' of their relationship with the vulnerable man living upstairs. Exploring a home's pasts can be a way to access forms of sociality which move between everyday local 'micro-publics' (Valentine 2008) and more intimate and intense relationships with others. These can extend through stories of benign social exchange – of information, hospitality – in encountering previous residents. But the ability to share and come together can also be impeded by socioeconomic realities which act to isolate people. The relationship between same/self and different/other can be contingent upon wider socio-economic forces which shape access to forms of homemaking.

In the next part I will move to a focus on the inherited materialities of home. In Chapter 6 I will consider relationships with the home's objects, and then explore homemaking practices themselves – processes and rituals of renovation and modernisation which are more available to some than to others.

Note

1 Definitions of 'heritage' have tended to focus on the specific social and cultural practices, experiences and spaces relating to (albeit often competing versions of) a 'past' place. Rodney Harrison (2013: 227), as described, has criticised continuing assumptions about what constitutes ideas of 'heritage', pointing out that 'linear

time' is one principle of 'ordering and classification' used to 'manage uncertainty' and responsible for 'our modern conception of heritage, as salvage or preservation of that which is distant, old, hidden and hence authentic, opposed to the notion of heritage as a form of creative production involving the assembly and reassembly of things on the surface and in the present'. This reflects emerging approaches, for example focused on environmental concerns, which require exploration of anticipatory practices related to future change (May 2019; DeSilvey and Harrison 2019; DeSilvey *et al.* 2011). This chapter reflects such emerging ideas about 'heritage', within my focus on *domestic* experience and practice, in order to explore the impact of the more recent past and its bearings on the future (and indeed, one of the book's central aims, as described, is to extend the conventional focus on the 'pre-inhabited' as only pertaining to *obsolete* dwelling spaces, to include within an expanded definition places which have pasts but also *continue* to be inhabited).

References

Ahmed, S (2000): *Strange Encounters: Embodied Others in Post-Coloniality*. London: Routledge.

Blakely, H and Moles, K (2019): 'Everyday Practices of Memory: Authenticity, Value and the Gift'. *The Sociological Review* 67.3: 621–634.

Brooks, P (2007): *How to Research Your House*. Oxford: How To Books.

Bushell, P (1989) *Tracing the History of your House*. London: Pavilion Books.

Crang, M and Tolia-Kelly, D (2010): 'Nation, Race and Affect: Senses and Sensibilities at National Heritage Sites'. *Environment and Planning A* 42: 2315–2331.

DeSilvey, C and Harrison, R (2019): 'Anticipating Loss: Rethinking Endangerment in Heritage Futures'. *International Journal of Heritage Studies* 26.1: 1–7.

DeSilvey, C, Naylor, S and Sackett, C (2011) (eds.): *Anticipatory History*. Axminster: Uniformbooks.

Harrison, R (2013): *Heritage: Critical Approaches*. London and New York: Routledge.

Kean, H (2010): 'People, Historians, and Public History: Demystifying the Process of History Making'. *The Public Historian* 32.3: 25–38.

Mauss, M (1990 [1925]): *The Gift: The Form and Reason for Exchange in Archaic Societies*. New York: Norton Press.

May, S (2019): 'Heritage, Endangerment and Participation: Alternative Futures in the Lake District'. *International Journal of Heritage Studies* 26.1: 71–86.

Valentine, G (2008): 'Living with Difference: Reflections on Geographies of Encounter'. *Progress in Human Geography* 32.3: 323–337.

Part III

Material pasts at home

6 Found objects

The tangible past at home

The poet Simon Armitage, reflecting on a stay in the spare room of a house during a long-distance walk, observed how such rooms are

> nearly always reliquaries or shrines, museums of past lives or mausoleums devoted to a particular absence, a place of mothballed clothes, stockpiled books, musical instruments, locked in cases, photographs under cellophane, framed certificates, dusty trophies, threadbare soft toys, objects which have no function or place in the everyday world of the living room or the kitchen … but whose significance to family lore borders on the sacred. I am sleeping in a memory vault and none of the memories are mine.
>
> (2012: 174–175)

As a corollary to complaints about the rubbish left behind by previous residents, participants often ruefully passed a closed door, requesting I didn't look in at that particular room. Like the room Armitage had stayed in, these were ad hoc memory vaults, archives of things which did not fit the needs of everyday life and were therefore disparaged as clutter. Jane and Nigel said their son's old bedroom was a 'mess'; Martin and Carol had a 'junk' room. Gemma described hers as the 'Room of Doom', containing objects 'in transition' (she didn't know where they were going; they had been there long enough to be granted the status of 'transitional objects', in a state of perpetual anticipation). For those in long-standing family homes, such objects proliferated – there was no compulsion for them to be moved, they just got passed on to the next generation – houses 'full of things' (Bella), just 'stored away' (Rita). The large archival spaces of such homes became emblematic of family continuity, and thus granted benign value. Rita described the 'magic' of the attic, where objects had been stored for a hundred years. Such spaces, behind closed doors, became unseen depositories of people's *own* memories, additional to the layers of material past left by *others* (Bachelard 1994 [1958]). For some, these archived objects of selves and others became entangled. Chris and Josie kept found objects in little boxes, alongside boxes of other inherited *family* items which included 'just little things you would lose otherwise … bits

of paper the children wrote "I love you" on. This is an old printer's tray ...
That was my mum's brooch' (Josie). For many, photographic archives, neatly
displayed in folders, or falling out of plastic bags waiting to be sorted, showed
the evolving biography of the home as backdrop to social events and meaningful gatherings, but also as *foreground*, the *befores*, *durings* and *afters* of
renovation work – people smiling atop ladders, covered in paint with brush in
hand; the physical home either centre or periphery, but always *there*.

Hidden personal histories

Burials, with or without memorial stones, are another way personal and home
histories become entangled. Chris and Josie, for example, pointed to a line of
three gravestones in the back garden, dating back to the 1940s, close to the
side wall of the house. In contrast to the many animal bones found in the
garden which 'didn't get gravestones' – including, as Chris noted, 'at least
one quite big dog in a box' – the cats were assumed to be Miss H's domestic
pets rather than rescued animals. The stones, names and dates indicated an
emotional response to loss. Deciding to continue the line, Chris and Josie had
added their own cat's grave: 'we buried our old cat in the fourth one along ...
in the next spot ... We've carried on the tradition' (Josie). Theirs doesn't have
a headstone – Josie thought this to be 'too *gruesome*'. Their cat, she joked,
was 'the unknown soldier', but not something they needed to publically mark.
They knew it was there, it could remain hidden.[1]

Ben in south London described a similar private burial, this time of a placenta after the birth of a child, over which they had planted a tree during a
family ceremony. He had later uprooted the tree and replanted it in the next
garden after moving:

> The big tree, this one here, that's exactly the same age as my son, who's 14.
> And we planted that tree for him at a birthing ceremony. And we kept – it
> sounds a bit weird, but we kept his placenta and we buried it under the
> tree. ... We planted that tree in the garden there, and we brought it with
> us. ... And that was the one thing I took out the garden – that tree, which
> was only a sapling. ... It's the one thing I didn't want to leave.

Ben joked that he would have to 'cut it down, turn it into logs' if it were to
travel with him if he moved again. He had accepted that the tree was now
too big and would have to stay – its unmarked, uncommunicated story and
personal meaning lost on future residents, a reminder of what of necessity
gets left behind due to the transient nature of home-making.

Inherited objects

It was precisely the fact that people could *not* fully know the original meaning
or emotional value of most things they inherited when moving into a

home – when entering the 'memory vault' full of *other* people's memories, as Armitage put it – that such left-behind objects required 'placing', needed to be handled with care in order to reflect their imagined past value. This value, decided by the particulars of form and geography and social context, dictated how found things were handled. As described, the more fixed and original elements of homes were afforded a high value for the benign sense of continuity and anchorage they offered, whilst others – such as inherited beds or baths – could unnerve and therefore were perceived more negatively. The decision to keep or remove objects was often dictated by the assumed meanings they reflected in relation to what they *communicated* about a particular past – often in the form of *stories* they gave life to or suggested. As Jack in Bristol explained about 'old stuff': 'I love anything with history that just looks out at you'. Adam made a similar point: 'Always a good story is worth preserving. I always used to collect things when I was a kid. I had a museum in my bedroom, you know. I like the stories'. Nigel recalled how they had wanted to replace the glass in a front window, but changed their mind when they were told its story: 'It's war-time glass, when you get glass that's got fluctuations in it and you can see it distorts ... The first V2 exploded near here. It told a story'.

For most, inherited objects were assumed to reveal the secrets of the house itself, particularly those which had been discovered during renovation work – the home becoming, as Leah said, an archaeological site. The objects on the home's surfaces contrasted with the hidden spaces or things which emerged over time – sometimes this involved actual hidden spaces, such as Jack's concern about something 'behind the wall', the unseen cavity also reflecting his emotional response towards his new home. In contrast, Penny told a more positive story of discovering a blocked off section of a bedroom during renovation work – here it becomes a *bonus* space:

> Because we've had to look at the house much more closely and we've seen what lies behind the plaster and inside the cupboards and under the floors ... the actual fabric of the walls. ... Oh and it was very exciting. [Our son's] room upstairs – quite a small room and one of the things we chose to do was to make that a little bit bigger. And there was a cold water tank. So that was going to make a little more room. ... Water tank came out. And then on the other side of the partition – *another* chunk of room. ... Jacob had a kind of vague idea that there was something that wasn't accounted for. And now there it is.

Small found objects

The shear range and number of objects discovered – especially the smaller, mobile things apt to get lost in a home's crevices – suggests homes are also containers and depositories of *various* pasts, a multi-layered archive revealed through the act of living within its spaces. Yolanda kept door hinges from the original doors and, in an envelope, flecks of old paint – the part standing for the whole. Brenda described finding 'bits of coal' beneath a hole in the cellar. Participants in the larger houses found bell pulls used to summon servants. Leah described finding these under the stairs which she left there, joking: 'Let's get these reinstated so I can have a cup of tea in bed'.

Chris and Josie opened boxes to reveal a surfeit of finds – a small lead battleship, a carved wooden pig from under one of the top bedroom floorboards – previously a children's room or school dormitory – a tooth-brush made out of bone, a 'little porcelain something-or-other', a squashed thimble, little dyed counters from Snakes & Ladders, glass and terracotta marbles, a door knob, bits of a China doll's face, buttons, a little cock dug up in the front garden, some old metal spoons from under the floor, a Victorian Father Christmas, possibly cut out from a card found 'in the loft, in the servants' quarters', a 1930s-style dress clip in the shape of a swallow, and a selection of old coins ('That's Queen Victoria's head. That's George the Sixth. That's Queen Victoria. I don't know about that one. Some of them are ha'pennies').

As described, there was also the inheritance of more porous substances, such as fine dust, soot, stains from cigarette tar, and some participants worried about having inherited contaminants, speculating on the previous use of materials such as arsenic in paint or wallpaper – a particular focus of anx-ieties about the presence of the past.

On the wall

Old wallpaper is another ubiquitous find. As described, Karen discovered layers of wallpaper stuck on top of each other in a previous home. Chris and Josie had kept a roll of wallpaper which they unfurled from a box, covering one end of the dining room to the other. Lydia had kept a piece of old wallpaper and placed it in a drawer as lining paper: 'I kept this bit just to remind me'. Brenda described finding old wallpaper lining the back of the cellar door, which she has also kept: 'I think it might be '30s. You see it could be something older or it could be '70s pastiche. I'm not sure which it is. They *look* to me to be quite old'. Leah and Alex also discovered wall-paper in a high built-in cupboard, now kept in a scrapbook of 'random' things: 'It was definitely Victorian because it was hand painted. It was like this greyish-greeny colour with white and red painted flower patterns on'. Yolanda described how, when she stripped the walls to redecorate, she found the 'original instructions, written in pencil, on the walls. So it'll say for instance, "cream", so that must have been the paint colour. And then a number, which I assume was a wallpaper reference number'. She 'preserved' the finds with photography.

Elsewhere, walls were also depositories of words or marks which offered clues about particular lives. William described chalk marks on the wall out-side his council flat, still faint traces visible, telling the story of the children who had previously lived in his studio flat. A neighbour told him they had only been allowed to play in the cramped corridor:

> The previous kids drew on the walls – as kids do. Squiggles. There were no actual faces or anything, they were only little children. There were lots of circles. I wondered if this was the way they released their *energy* because they were never taken to the park. I didn't take the chalk marks off for a long time.

Chris and Josie had two significant discoveries on walls. Firstly, the sketch of a man's head on the wall of an upper bedroom, which they traced before covering with wallpaper. This was a 'caricature ... a profile of a man, fairly elderly gentleman, with a little goatee beard ... drawn in pencil on the wall' (Josie). They assumed this to have been 'drawn by one of the girls', during the time the house was a boarding school, or by other children, or indeed, by a decorator: 'I mean there's speculation'.

Up narrow stairs above the kitchen is a storage loft where they found on a wall lists of apples by type, weight and year – mainly from the 1940s and early 1950s: '12 pounds of plum, 12 pounds of this, that and the other, 1942, 1943 ... So obviously that was their jam store', part of the 'war-time spirit ... producing what you can with the resources that were available' (Chris). They painted around the lists but left them as they were 'part of the history of the house'. Like continuing the line of cat's graves, they carried on the jam tradition, the 'little ancient trees around the walls' continuing to produce fruit. Chris had just made a batch:

> Most of it just goes in the freezer and comes out for puddings. The apples I religiously pick every single one. Can't waste it ... And the children don't eat them because – they might have a *maggot* in them. They prefer the supermarket ones. So I eat most of the apples as well.

But Josie insisted this was not a sentimental activity undertaken because 'somebody else has done it'. She explained: 'I'm doing it because the plums are there. Presumably that's the reason *they* did it, because the plums were there. You don't want them to go to waste. You've got to find some way of preserving them'.

Elsewhere, Anna in east London described finding a Jewish *Mazuzah* – an encasement attached to the frame of a door containing a prayer on a small scroll. From this she gathered previous residents had been Jewish; this was confirmed when she traced a Jewish community, since moved away, which had lived locally during the mid-twentieth century. She checked if the prayer was still inside the casement:

> I was curious to see if it was there. You know. And so I carefully got the paint off all round it ... It was about, 'I wonder who's this was?' ... And I think [lowers voice] the builder must have thrown it out. Because I've

lost it. I realised it's not there. And I don't know when it disappeared, but I just suddenly realised it's not there anymore. I'm disappointed. Because that was a little bit of – *history* – that belonged to somebody else.

Her frustration at the object's disappearance reinforced her belief that there was an obligation to keep it – it was not hers to lose. Like the jam archive, it was a part of the house.

Joy in her east London rented flat, pointed out a different, more recent, form of wall inheritance – the letter 'E' nailed to a bedroom door:

I still never found out who put the E on that door. I've never tried to take it off. Because – I laughed, when I come in and I saw it ... I said to my son, 'Look, mum's moved with us' [laughs]. My mum's name was E ... I don't know who put it up there or why [laughs]. But that's *mum* [laughs] ... It was just one of those instant things ... It was just that instinct, you know. You see something and you just say it. But it *stuck* ... As I say, I don't know who even put it up there [laughs]. But it still connects with [my life].

She knew rationally that the letter would have been the initial of a name marking the territory of a bedroom, probably that of a child, but she preferred to believe it was a sign from her deceased mother. Whereas for most, wall markings were further clues to previous lives in the home, Joy wished to personalise the find, creating a different meaningful connection.

Natural inheritance

Apart from the possibility of unwelcome left-overs from the past such as arsenic, and the bones of pets in the garden, participants also reported other forms of natural inheritance. Martin found the skull of a pigeon in the ceiling which had 'obviously got stuck up there', and 'inherited' an obligation to look after some cats: 'There are about three feral cats that this guy used to feed. And I'd been heartless and said, "Oh well, that's it". And Carol said, "No. Now we will feed them". So Carol does'.

Elsewhere, a desiccated rat fell on Chris whilst working in the cellar: 'Old plasterwork, yeah. Nasties under the floorboards. Very ancient and very dead so yeah [laughter] ... Just normal casualties'. They also found a nest of butterflies in an upstairs shower room when they moved in which, 20 years later, had continued to be a spring-time event. Like Anna's repeating plant, their reappearance became a 'faff', a 'nuisance', particularly as they had to be cleared out before taking a shower:

CHRIS: Years and years and years and years afterwards, this room was full of tortoise shell butterflies every spring. I think they just somehow sort of nested ... We'd just put them out of the window.

JOSIE: But even now in the spring I sometimes find them in the corner, in the shower … I think they return to the same spot.

A further story told of an apparent repeat infestation reflected attitudes to an outside space. Harvey and Rochelle described how the garden had been a working space and a 'privy':

HARVEY: I don't think that the garden was *ever* used as a garden by anybody. When we got it, it was just – *turf*, just raw turf. And it had fox holes in it.
ROCHELLE: [emphatic] They were *poor* people. In fact, we wondered at one stage whether it was [very quiet] an undertaker's, actually. There were people who were cabinet makers. And I mean, the garden must have been used as a privy. You get a lot of flies there. And I'm sorry –
CL: Even *now*?
HARVEY: Even *now*, yes … But I mean, it would have been a *yard*, really – a work yard, whatever work they were doing.

The confusing description of the continuing nuisance of the flies is one of the justifications the couple made for making radical changes to the garden – it was not really a garden in the first place, nothing to be valued as 'history'.

Old trees in gardens were more valued, but not if they were too close to the house. Brian, for example, took down three old trees which took up too much light. Megan pointed to a small hawthorn which she believed had self-seeded from one of the original ones. She took pleasure in that – the older tree had been regenerated, found new life.

Digging up history

Apart from the animal bones, there were many stories of interesting finds from the garden, a site of domestic archaeology (Owens *et al.* 2010). Brian and Megan told a hopeful anecdote about a neighbour finding buried treasure – a James I gold coin dating to 1684 that was in such 'perfect' condition that, as Megan put it, the 'eyelashes were completely decipherable'. The coin – which eventually sold at Sotheby's auction house – had been found when digging the footings for a greenhouse. The neighbour had popped it in his pocket thinking it was a 'rubbishy old penny in the soil. At the end of the day when he went upstairs to have his bath, he took it out and scrubbed [the soil] off'. The couple imagined how the coin had been originally dropped, favouring a scenario where it fell out of the pocket of a man riding a horse across the fields. The excitement of the find led their sons to borrow a metal detector – without success: 'They found all kinds of bits of rubbish – wire and coat hangers' (Brian).

Elsewhere, treasures were unearthed by default of the movement of the street. Lillian, living on the major south London road, described how her long front drive, before it was tarmacked, would 'throw up things' to the surface due to the movement of the road – a gold ring, coins, bones, 16th and 17th century pottery, clay pipes, 'all sorts of bits and pieces':

> You'd go out and you'd suddenly see a ring [after] a rainfall ... or a lorry going by and then the earth moving ... This house has no foundations. So underneath the floorboards, underneath here are pyramids of brick ... The house moves. If you have a lorry going by, you know, a really heavy lorry – it does move, it does shudder and shake ... I've got insurance if it falls down [laughter].

The afterlife of objects

Adam described how, when walking along Hadrian's Wall, local people told him that 'if they dig up Roman relics in the garden, they just bury them again because they don't want all the bloody tourists and all the archaeologists coming round. They just bury them again, you know'. As described, the *circumstances* around the discovery of found objects in homes – those attached to the surfaces of the home and those hidden and revealed through renovation or garden work – became important in dictating *how* they were handled, the decisions made about what to do with them. The manner in which this was done revealed attitudes about the pasts they represented and their situated contexts – what stories they told, the particular values granted – aesthetic, monetary, historical, imaginative, symbolic – as well as reflecting or changing participants' relationships to their homes. The remains of buried dead animals were left in situ or, where necessary, disposed of, whilst those which returned – such as butterflies, flies and plants – needed to be handled differently. Pencil or chalk marks were also removed or accommodated, preserved through photography and tracings, painted around or covered.

Recycling objects

Some found or inherited things with practical value were also recycled. Unlike the lucky gold coin find, few objects were deemed of monetary value, although Leah found a silver bracelet and Lillian a 'lovely golden turquoise' 14 carat gold ring – one of the objects 'thrown up' from under her front drive. She gifted this to 'an Iranian friend of mine because she loves turquoise and gold. And she wears it a lot'.

Other items, such as horseshoes, were used as ornaments. Broken china pottery from gardens was also recycled – Cathy placed pieces at the bottom of plant pots for drainage – as well as creatively refashioned. Chris and Josie showed a mosaic created out of pottery pieces which hung in their kitchen. Gemma also made a mosaic, cemented into the garden concrete, the smaller fragments used to spell out the year it was created. Elsewhere, Leah and Alex described finding a good quality Victorian chisel 'from the time the house was built', and enjoyed the visceral frisson of reusing it. The chisel had been personalised with an initial – a 'W' or 'M'. They decided on 'W' and speculated on the name – William or Walter:

> I found it tucked under the floorboards. ... We found lots of old newspapers under there and stuff like that ... must have been when the house was upgraded, the plumbing was upgraded, it was 1910 a lot of it. ... But this is older than that from the style of it. ... It's a perfectly good chisel. This is for chiselling-out door locks and things like that, where you have to chop into the doors. ... I've used it for chiselling-out doors. ... It's a nice feeling ... that you're carrying on using a tool that someone, you know, 150 years ago made money from, if you like.

Alex was drawn to reflect on the better quality of past workmanship ('they definitely did make things better then. There was a lot more care taken to produce it ... These people had to go through five year apprenticeships to learn to use *that*'). In turn, having established it had been used by a 'proper craftsman', thoughts turned to why he left it behind. This, Alex believed, would have been a 'major mistake'. It 'might have cost him his job, his life. He could have ended up in the debtor's prison [laughter]'. Although there was concern about the plight of the chisel's owner, there was a flippancy in the speculation. The reuse of the chisel, like other continuities, created a 'nice feeling' of symmetry, more important than any dis-ease created by thoughts of the previous owner's plight.

There are other examples of the reuse of past objects. In Anna's house the end of an obsolete gas pipe was sticking out of a wall of a bedroom, painted over so many times to be rendered smooth and blunt. The pipe end lent itself to use as a hook: 'It was obviously too much to take out. We could saw it off. But [we] find it quite *useful* for hanging a dressing gown'. The pipe end became part of an embodied, practical, intimate engagement with the lingering, obsolete traces of the home's functional elements, breaking out here from their hidden spaces onto the surface.

Lydia, in her ex-co-operative tenancy, pointed to inherited things she kept: a chest of drawers, a cupboard on a landing, a red carpet on the stairs, and a large dresser, which she had painted white. She explained: 'It started to contain all my junk. Its *use* became more valuable than what I considered its ugliness'. She made a connection between her need to reuse what she inherited due to lack of money, and a feeling of transience: 'I'm a make do and mend person because I've never had all that much. I do recycle. ... There is another attitude as well which is always, 'I might be moving out some time'.

The dresser also reflected an ethos of collective sharing and recycling as people moved in and out of different co-op houses. Lydia recalled someone visiting from a flat around the corner who told her:

> 'Do you know, that used to be *mine*?' I don't know how it got from her house to this house ... A lot of shifting houses has gone on. And then a lot of shifting out. And shifting of furniture.

In different circumstances, Sandra, in the charismatic crumbling house next to the motorway, also 'recycled' objects from other flats in the house. These were salvaged from the abandoned possessions of tenants who left. She had observed them leaving on numerous occasions – as described, at ten years she was the longest tenant in a transient population. In many cases people left in a hurry, abandoning usable things which were dumped in the front of the house or loosely placed into untied bin bags. Unwanted items, such as clothes and books, she took to charity shops, but she showed me her own gains: a plastic stand-alone CD rack in a corner of the living room, a set of matching rice dishes in the kitchen, a painting of the reflection of the moon on water hanging on a wall. She had not liked the painting at first, using it to cover a stain, but it had grown on her:

> I mean, I wouldn't have bought this if I saw it in a shop. It isn't really my style, but it hides a water stain. I like it now, I've got *fond* of it. I think it's peaceful, which belies the *lack* of peace here. It actually was still in the *wrapping*.

The items hinted at untold stories, and became part of Sandra's social commentary:

> Whenever someone goes, they gut the place. So because people have come and gone *so often* – you come back home, it's the usual scenario of the mattress by the bin, and the piles of what are people's *effects* – and things that they obviously should have taken with them, like books, pots and pans. And you think, 'Well, this is not right'. Because when people leave, you don't leave that stuff behind … Either they are leaving in a hurry, or she's chucking them out. There's something dodgy about it. And it happens so many times. … Heaps of times … I think some of them must have done a bunk because there were perhaps court proceedings or something. … A court day was looming and they just decided to – *bugger off*. I don't know. It is weird. Perhaps someone else was after them, like immigration. I know she had loads of asylum seekers, some of whom paid and some who didn't. … Some of them had to go so quick that they have literally – left everything behind.

Accidental preservation

The reasons why these objects were left behind may never be known, but as Sandra suggested, they became the residues or by-products of chaotic lives caught up in wider dramas. More broadly, people speculated about the reason why some objects were left behind – what was intentional and what accidental – and this became important in 'placing' and assessing the meaning of the objects as they emerged in the present. Sometimes the reasons for being left seemed obvious, such as small objects lost in cracks or thrown away in gardens. Susan found an 'enormous' metal trunk in the attic inscribed with the surname of the previous owner, an older woman who lived locally. She asked her:

> 'Do you want it back?' And she said, 'No'. I said, 'Because we can't get it out the attic as it is. I don't know how you got it in there!' [Laughter] It's bigger than the hatch. And she said, 'I don't know how we got it up there, but it was up there and it stayed up there'. I said, 'It is still there and it's clearly yours' ... It was like the kind of thing you can imagine some child in the '30s going to boarding school and, you know, with their name on the front ... they'd have their clothes and their tuck box, I suppose, in there. I can't think what else it was used for. I should ask her.

The story of the trunk is one of *inadvertent* inheritance of an object that got stuck – no one knew how it got there, but the hatch space was too small to get it down.

Jack in Bristol described finding old bone-handled knives and forks and a carving knife with a date on it – 'a wicked looking thing' – inside a jammed-up drawer which, being hidden, he speculated, had been forgotten, ignored or just failed to be opened by the previous owner. The cutlery could be separated from his negative feelings about the woman because of their quality and historical value: 'I can *disassociate* ... I appreciate the quality. I mean ... it's got a bit of history behind it'. Despite feeling the need to emotionally and materially 'purge' the vendor's presence from the house, he was happy to keep these objects. They had histories which predated her.

Many objects were assumed to have been 'preserved' by accident because it was cheaper or easier to do so rather than because of a *desire* to preserve them – indeed, here it was the *lack* of care or value placed in objects which ensured they survived. Sandra made this point about the strange exterior features of her house which were 'still there – probably because I'm not sure she [the landlady] cares less about them, one way or another. By default they're still here'. Alice in Hackney pointed to the original shutters, noting that 'no one really cared too much about them' so they didn't notice them enough to take them out, whilst Alex made a similar point in relation to the original wallpaper: 'they didn't take time in cupboards'. It was assumed that previous residents felt no need to make the effort to remove or paint over the residue of the past found in storage spaces.

Many participants commented on their awareness of how current interest in the past contrasted with the active *dislike* of inherited features during the modernising 1950s and 1960s. Karen described how

> DIY programmes were teaching ... how to put a bit of wood in to flatten off, get rid of that *lumpy bit* [laughs]. People were all putting bits of hardboard over the front of the fireplace and wallpapering over the whole thing – to get rid of it.

Although people stripped out and replaced features, it was often more expedient to cover them up, as Leah added: 'The intention was probably "it's much easier just to put a board up and we didn't have to think about it"' – an accidental form of preservation.

Some participants reflected on broader historical *happenchance*, referring to older, more structural features left intact because, again, it was cheaper to do so – a version of 'make do and mend' on a larger scale. This was Bella and Sylvia's argument, in the old manor house near Bristol. They pointed to large stone slabs over fireplaces which had been taken from the remains of a Roman settlement in the grounds – an early form of salvage and recycling. Other objects had remained by chance as their house had been rented out by its aristocratic owners for long periods of time. The landed classes would inherit houses but rarely sold them on, they explained, unless (as had happened once during the history of the house), there was a bankruptcy. Such houses instead were often let, sometimes to poor relations – in their case, specifically, for 200 years from 1420 and later again for 250 years. When letting a property, Sylvia noted, the owners tend to do the minimum maintenance necessary ('the picture here shows what a bad state of repair it was in in 1880'):

> And that's why it's interesting. Because people do not remove, you know – put fresh things in – they do what they have to, to let it ... All those periods of time, people just did what they had to do [which] preserved the interesting things.

In turn, she mused, aspects of the house that got covered up eventually – such as the Roman well which the Victorians had 'flat stoned over', had just 'got forgotten. Things do like that'. Bella pointed to the blocked-up staircase, which the family had lived with unknowingly for decades: 'No one even thought about it'.

Hiding things might also be a more intentional strategy for preserving them, but Bella and Sylvia were annoyed by what they considered an illogical response to the question of what should or should not be *uncovered*. Railing against strictures imposed on their home as a listed ancient monument, they explained that they had been refused permission to unearth the Roman settlement in the grounds, traces of which could be found in the shape of the land:

BELLA: [Sighs] We're not having a dig, because we might not leave anything for anyone else to uncover another day ... It's about leaving something for other people.
SYLVIA: It means that if you don't uncover it, you have left it still not spoilt.
BELLA: No, well you aren't necessarily going to spoil it by digging it up.

This obligation to preserve the archaeological remains by keeping them covered up was considered an unnecessary attempt to *control* the future of things, in contrast with the reality of the ad hoc process of how things often got preserved through being recycled for other uses, or through being accidentally or expediently left alone until they are forgotten and then, perhaps, found by chance later on.

Leaving a mark for the future

A different form of material inheritance embodied a *desire* to be found: names, initials, dates scratched onto walls. These were assumed to have been intentionally placed by a builder or resident for no other purpose than to say, 'this is me, here, now', 'this is where we live, this is our home', or 'this is what we have done to improve it'. Such scratchings mark effervescent moments of pride and senses of embodiment in a place; abbreviated shout-outs to future residents, calling to be remembered by imagined others. Usually these appear in materials and spaces most likely to last or of particular value. Bella and Sylvia pointed to the initials and date on a beautiful, carved wooden staircase:

> I do think about the people who put the new staircase in, the 1683 staircase and all that. Because I reckon that they were people who loved the house. ... We had some architects round and they said it's very unusual to get a hanging staircase in a house like this, so they were obviously doing something quite dramatic at the time, these people.
>
> (Sylvia)

The house offered insights into its evolution to a modern home, the shift in sensibility mapped onto its physical development. They identified more with previous residents who had put effort into care for and improve the home, sharing their emotional connection and pride by 'signing' the staircase. These initials and date would not be necessary if a home was purely a 'shelter', Bella reflected: 'When did they start feeling a sense of *place*, if you like? A sense of – *more than* just "this is a good place to live"? But – "let's have a house that's worthy of where we live"?'

Others also found names and dates. Jacob and Penny found names scratched onto an upper window of their house, and Gemma also pointed to initials on a stone slab on a side wall of her cottage. But as her own mosaic date mark suggested, some participants also engaged in naming and dating rituals themselves, sometimes in order to communicate something specific to future residents. Chris and Josie showed me a fireplace in a style original to the age of the house which they had bought at an auction. With an archivist's sense of responsibility to record its provenance, they had become concerned that future residents should not be duped into believing it was original. Josie explained:

> I put a note cemented in behind the fireplace stating when we bought it, and what we paid for it and the circumstances ... [so] they will know that the fireplace isn't original to the house ... if anybody ever takes that

fireplace out. They'll know that it wasn't put there in 1805 which they might assume ... I want them to know the fact that we bought it and put it in in 2006 or whenever it was.

For Josie, this was a professional instinct, but also a personal statement. It is, she said, 'putting us in it. I'm pretty sure I put our names ... and the date we bought it and how much we paid for it'. The personal additions are less necessary than the fireplace's own biography, but she wanted to 'place' herself in the history along with offering information, conscious also of the possibility of sharing the pleasure *she* felt in discovering hidden things: 'How lovely it is to find these things. I suppose we quite like that idea ... It's part of a tradition'. The irony – that in order to find the note the fireplace would need to be removed to be discarded or replaced – was also not lost: 'It's a contradiction really, isn't it? The people who rip it out will be sort of not interested [laughter]. Otherwise they wouldn't touch it' (Chris).

Others also enjoyed carrying out playful versions of this tradition – one which linked to the craze for time capsules, and, earlier, foundation stones and other protective rituals. Anna, for example, wrote her builder's name under paving stones, and, while helping a friend wallpaper a room, they wrote their names and the date, 'so that when people strip it off, they'll see it's there'. These are unobtrusive ways that people 'add' themselves – and their builders – to the layers of buildings. Being hidden, they don't take up space in the *present*. The house's layers offer leeway to explore paradoxical senses of belonging and identity which acknowledge temporariness as well as the possibility of continuing memories. In this way, the markings are a kind of memorial of themselves, only available for the future when their own time will be over. As fragments of communication across time, these gestures reverse the frisson of imagining how found objects were touched by *past* residents. Lillian's take on this reinterpreted 'tradition' comes in the form of placing 'wishing well' deposits of coins under floorboards. She explained:

When we moved in I did get some coins of the time and put them under the floorboards. The coins of the year that we moved in – 2001. Several, you know, pennies – two pennies and one pennies. And they were all under the floorboards. And I've done the same in my [other] house as well when we moved in. I put some under the floorboards. So that when people restore it they can find – 'Oh, there's a coin!' [Laughs] ... It's leaving a mark ... It's just something I wanted to do. It's just for fun really ... And if it's ever knocked down, they'll say, 'Oh look somebody' – you know ... I recently had a carpet put in, a stair carpet. And I said, 'Look, could you put this little coin in'. And the builder said, 'Oh, a lot of people do that'.

The 'tradition' often marks a moment of change – partly because the opportunity arises most when layers of the home are temporarily removed, allowing space to hide things before new layers are added. It is also a tradition of the *builders* who do the work. Whereas residents tend to assume communication with future residents, builders do this to communicate with future builders. These rituals were revealed in an interview with Peter, a carpenter, who explained, firstly, that it was a way of communicating *practical* information to the next generation:

> I've left things behind knowing it would be useful if anyone came after me to know when I did it because of the material science of the time. So if I said this was done in 2002, as soon as they find that paper they don't need to question whether there is any asbestos in that job ... or if they are going to find any weird screws in weird places. Because if I made it at *this* time I'm following *these* procedures of building, whereas if I made it at *this* time I'm following *these* procedures, or those materials, or those fixings. I have personally done that, where I felt it would be useful for anyone coming to that property afterwards.

Builders often leave *newspaper*, he added, signalling the different intentions in doing so through the way this is presented. A 'bit of scrunched up' newspaper is a 'standard trick if you've got a big hole ... to fill the hole with, and then you put the filler on top of that. Now we've got expandable foam instead. But it used to be newspaper'. In contrast, a *neatly folded* newspaper was 'put there intentionally'. Peter explained:

> Sometimes you'll find just the first page or the last page of a newspaper. When builders do it, they do it with the intention that this is to be found. So they don't want it to look like rubbish. [Gestures towards a wall] 'Ok, I'll put that in there. Lovely'. Then they go home and tell their missus and they forget about it [laughs]. And it's just one of those things that people do ... If you find initials, or a name, it is often just pride. Or, 'Ah, I'll put the newspaper in so that whoever finds it later can have a bit of fun' ... When guys put in newspapers in 1910 they must have been doing building work on older buildings and have found similar stuff themselves. It's a long-standing tradition to leave things in walls, really. Builders do it a lot – they've always done it. And because you find it in your trade – as you build you find these things – you go, 'Ah', and you do it yourself. It's like a little storytelling tradition. Nobody ever tells you that this is going on. You just find something and that makes you think, 'Ah, I'll do that myself. I quite like that' ... Everyone knows it happens ... because I've talked about it when I've found stuff. 'Oh, I've found this and I've found that'. And people start talking about what they've found over the years. And – then they start talking about what they have left. Because it's just – part of what we do. You know?

This oral and embodied tradition also allows builders – often ignored and rarely *named* – to acknowledge each other's work:

There is a pleasure, there is a *genuine pleasure* in finding evidence of another man's work, and knowing that man's dead – he probably had a long career and ... helped build x number of houses in the Victorian stock – or it's a later date and you know that it's a work of rejuvenation ... *Pride*. Because why would you leave a note with your name and the date or whatever if it wasn't to say 'I did this, at this time. And I feel a *value* attached to that. Beyond just the money that I'm receiving'. Builders don't get recognised. This is just a house that some builder's built. As soon as you see 'Dave Jones' on the wall, inside something, you think, 'Ah, Dave Jones built this' ... It's kind of exciting. You're connecting yourself to the future ... *Another* builder might find it. And another builder. The only person I care about really is another builder's point of view ... I feel genuinely like someone's given me a compliment if a builder who knows what he's *talking* about and knows how to do what I'm doing, comes in and says, 'This is good work'. So even though you're not going to *receive* the compliment, the idea that a builder in the future will open up what you've done, find your name, and compliment you in his own mind.

Rituals of reburial

The range of rituals and 'traditions' to be found involving hiding things suggests a desire to be part of an *idea* of a tradition (if creatively interpreted through mutating forms; see Blank and Howard 2013) as a way to connect with others across time in benign, fun, unimposing and even practically helpful ways. These also reflect different relationships to the material surfaces and depths of the home itself, here in creating conduits between past and future residents and builders. The depths – containing things not on show or not intended to be too easily found – become domestic archives or depositories not just of accidentally lost or left-over objects but things left intentionally for the future, performing a desire to share an exclusive access to senses of home – an active form of *heimlich* akin to passing on secret recipes across family generations (Lipman 2019). Like the objects in spare rooms, these are liminal, waiting things embodying vicarious memories, holding patterns of shared belonging.

Covering things up with a view to the possibility of later rediscovery or reinstatement also keeps things intact and out of sight in a way which respects the home's *fabric*. Unlike Bella and Sylvia's frustration over their hidden Roman remains, the intention is often to preserve elements of the home one does *not* wish to look at. Harvey and Rochelle, for example, pointed out that the original floorboards were still 'underneath the floor'. Anna copied the front bay windows but placed the original ones in the cellar for future residents:

If we *sell* it, we'd say to the people – you know, 'They are *there*. You can have them or not'. I think they're just so lovely that I don't think we should [laughs] – I couldn't have, I just couldn't throw them out.

Elsewhere, other participants emphasised how the changes they made to their homes were *reversible*. They wished to stress their sensitivity to the home's original fabric. Even though covering things up that were no longer fit for purpose or aesthetically appealing related to similar post-war era practices (assumed to reflect a desire *not* to dwell in the past), participants consciously distanced themselves from drawing too many parallels, stressing that their *intention* in doing so was about *respecting* the home and its past and future residents rather than merely acting out of expedience. Brenda, for example, blocked off a door to create a wall for kitchen units, but only used plaster-board: 'What I've had done here has meant it can be returned to what it was earlier if necessary. Moving the kitchen to another room – future residents can *reinstate* it'. Lillian also described having to take down a broken wall but she 'kept the pieces ... It's only small pieces I've kept but I've got a sort of section'. She also placed drop ceilings underneath the original ones to cover them up, and painted over glazed tiles around a Victorian fireplace, emphasising they were still intact:

> I wanted just plain ... I mean none of it is damaged. It's all behind there. They're not damaged. You could scrape this off – I've just painted over them ... You know, this is *my* taste. Somebody else might like them. So I didn't want to destroy them ... When we go next door, I've blocked up a fireplace. But I've kept it there. It's still there. But it's blocked up. Nothing is *gone*. It's for the future. I'm only a guardian.

Double doors between two reception rooms were also plastered over but maintained: 'Behind here are double doors. I've kept them. And I've got pictures of them. You can see the hinge, you see? ... They're still there. I just took the handles off. The handles were modern anyway'.

Practices of hiding or burying things also shed light on what people do with the smaller found objects. These were also subject to forms of 'archiving', squirrelled away into boxes, envelopes, scrapbooks cupboards, drawers, attics, cellars. Lillian described how she 'reburied' found objects, placing those things thrown up by her shaking driveway into a 'secret compartment', out of sight in a room full of glass cabinets displaying antiques, family heirlooms and gifts, each with interesting stories to tell – reflecting her previous career and travels as an antiques dealer. The found objects, in contrast, have no aesthetic value to her: 'They are in a secret compartment under there. There's no way I could display them'. The compartment proved so inaccessible behind the display cabinets that it remained 'secret', too difficult to get out. Hidden away out of sight, these objects have a different function. They are retained because they are a part of, they *belong* to, the house itself and its history. Indeed, all objects found by participants were kept within their homes. Broken pieces of garden pottery might have been subject to the most playful treatment because they were assumed to have been discarded – their *intention* was never to be preserved – but even *these* pieces are retained in some form within the home, often reused for practical purposes or creatively refashioned into art.

The *keeping* of things, then, signalled particular affective relationships to them. On the one hand, finding hidden objects felt like acts of intimacy between residents and homes, allowing people to feel they had been granted access to their home's past secrets. But these erstwhile hidden things – invisible, waiting presences of and from the past – are, once exposed, also granted agency. They needed to be placed into a *special category* of meaning beyond (even if influenced by) their perceived aesthetic, monetary or use values. The nature of these objects required a particular form of handling – perhaps more so if previous inhabitants were assumed to have lacked a sense of custodianship, or did not have the privilege, money or time to have such concerns. Either way, as part of the home's accumulated identity they needed to be retained *within* the home. The manner of their storage also contrasted with the way participants stored more personal items. By reburying found objects, participants were engaging in a ritual act – to respect the objects' histories and, by extension, the barely-accessible memories of past others. These practices became part of the negotiation of shared belonging, and a ritual of appeasement to the house; another important element of such rituals of reburial is to *contain* the past. Placing things into hidden storage containers and spaces out of sight literally kept the past 'in its place'. Doing so, as Lillian suggested, does not trouble the aesthetic and functional choices involved in present-day home-making. In this sense many found objects are granted *less* value. And yet, if the 'past is immanent in our embodied engagement with [the] world' (Harries 2017: 126), we also need to account for those objects which we do *not* wish to touch with hands or eyes, that we need to keep present and close, but simultaneously distant. This may be an attempt to diminish their agency, but keeping them out of sight within the home – like Jude Hill's (2007: 75) archived amulets – also granted them *more* power, reinforcing their mystery and acknowledging their special relationship to the home's past.

Keeping objects in situ

The concern to respect but contain found objects gave beliefs about the past's continuing atmospheres and energies a more tangible focus. Many participants reflected on earlier traditions of burying certain objects within the home which are often assumed to have been motivated by the desire for protection against evil spirits at geomantic weak spots (Billingsley *et al.* 2017; Hoggard 2019). Josie and Chris referenced their knowledge of this:

JOSIE: We know about people that find shoes in houses.

CHRIS: Things that have been lost or deliberately put into these hidden places.

JOSIE: Deliberately put or lost.

If deliberate, it is the *placement* of found objects within the home which becomes as important as keeping them *within* the home. This concern reflected anxieties about how to handle inherited objects correctly and the ramifications of not doing so. In the course of research, anecdotes abounded about edicts written into contracts about keeping, for example, violins in attics or other objects defined in relation to their *belonging* to a home or having become irrevocably *attached* to it. These beliefs create new rituals embodying particular obligations to act and some participants seemed surprisingly willing to accept such edicts passed down to them – invariably dictating a need to keep found objects in situ. Of importance here is the social context in which such rituals are enacted, in particular perceived hierarchies of authority; those considered most knowledgeable – including previous residents, older generations of family or local historical 'experts' – are granted particular influence. These relationships of power account for one of the ways such traditions get passed from one generation of residents to the next.

Objects with an assumed original ritual function – those which are considered to have been left *intentionally* – created the most anxiety, but the age, intention or function of such objects was not particularly important. One woman, Divia, wrote to me about finding a shoe in the attic of her seventeenth century house in a town near the English South Coast. She said: 'We took it down and then someone said that shoes are left in houses to bring luck and to protect the house. We put it back in a round balsa wood box that we had'. An assemblage or 'cluster' of other finds had been collected in the box since – an old photograph, a newspaper dated 1943 – but as she explained:

> The shoe has overarching significance. The other things gather and I think the shoe has to be there. Being asked by you about it, I feel nervous. I don't really want to even get it out. It feels embedded in the house. Part of its fabric and not something that should be meddled with ... Leave well alone ... We will leave it there and tell the new owners that it is there – hand it on as something to be *known about*.

Such rituals are not just confined to the older, more archetypally ritual objects. Elsewhere, for example, in an early nineteenth century house in south Manchester, the owners inherited a painting of a man, which they were told by the vendors was at least a hundred years old, came with the house and needed to stay where it was. They did not seem curious to know who he was, but they gave him a strategically light-hearted and unthreatening nickname, suggesting a desire to incorporate him into the family, to make him 'familiar': he was known as *Uncle Toby*.

A further example, from a previous interview, concerned a ritual passed down through the matriarchal line of a family. Susie, living in a 1930s semi in suburban Ipswich, had found a little plastic cat on a window sill when she moved in – the kind of small, cheap, mass-produced object which might have fallen out of a Christmas cracker. She explained: 'I don't move it. It stays

Figure 6.1 Fragment of a biblical quote ('the wages of sin is death, but the gift of god is eternal life') found on a wall near the front door of Pam's remote farmhouse; she believed this reflected the dale's previous Quakerism. The fragment was rescued during restoration work and is kept in a box.

there. … My mum always said: "If you find something when you move into your house, you must keep it. And if you leave the house, you should leave it there"'. She added that her mother had done the same thing, that it was 'something that she had maybe drummed into her when she was a child':

> There was a silk scarf, and an ornament. And [mum] wrapped the ornament in the silk scarf and put it carefully in the loft. She left it with a note to say that it must never leave the house … When we found the little cat – she said: 'Oh, you must leave that there. You mustn't get rid of that'.

Conclusion

Most inherited objects are retained within the home in rituals of respect, appeasement and (quite literally) containment of the past. To some extent, these practices around found objects suggest that people are *more* respectful or accepting of continuing presences than when dealing with living *people*. Objects speak back in rather more complex ways than recent residents, perceived not only as holding power for *once belonging* to others, but also – having a continual material presence within the interior – are considered part of the accumulative identity of the home itself. Hidden objects intimately revealed – be they in garden soils, behind wallpaper, under cracks in floorboards or in jammed-up drawers – thus become one oblique way the home reveals its story. The spontaneous discoveries combine the seen and unseen in domestic archives which shift between remembering and forgetting – the 'personal spaces' of collecting and display (Blakely and Moles 2019) becoming those of others as well as selves.

Specific ways of dealing with these objects depended, as explored, upon the various ways they were valued, as well as, in the above examples, sociocultural contexts around repeated practices connecting past, present and future residents within webs of relatedness and obligation. As explored in Chapter 3 in terms of beliefs about the 'presences' of the past, these objects are granted forms of agency which require they are kept within the spaces in which they are inherited, reinforcing essentialist notions about authenticity and identity abiding in fixed places of origin (Massey 1995). But there is a hierarchy of values which dictate emotional responses to a range of objects, often depending on the assumed intention involved in being left – whether hidden on purpose or lost by accident. Marks of habitation or work done – names, dates, newspapers – are also intentionally left in ad hoc oral traditions that continue to circulate locally, anecdotally. Although some participants reflect on their homes as archaeological sites, they are not hidebound by the rigorous collecting and categorising methods and the distanced, technical gaze central to conventional archaeological practice. This allows access to different forms of engagement with past objects – more affective, everyday heritage encounters often side-lined during professional practice.

Many encounters with the home's past thus involve revealing what is hidden and in turn, hiding what is revealed. There are other rituals, however, suggesting new relationships to the home being formed – and physically marked – as people live in them. These can be acts of burying things which once had or supported life (pets, placentas) – rarely signposted because the ritual is performed only to be privately shared; it does not need to be recognised or acknowledged by others. Leaving these behind can be a source of tension, reinforcing the complexity of rituals involving the marking of meaningful or impactful relations and events taking place during the process of habitation. This is one reason why people maintain an emotional connection, usually from afar, with past homes. These acts of burial also remind current residents that *past* residents may also have buried things ritualistically because they were meaningful – explaining why participants sometimes hesitate when choosing what to do with what they find; the meanings of such objects may be lost, but their sense of having once been meaning*ful* is not. Thus such domestic 'archiving' practices need to be understood in relation to emotional and affective relationships of care for objects which are not always *personally* meaningful to people, and the way people replicate – and are *conscious* of replicating – practices around objects which can create empathetic and/or anxious relationships between various selves and others across time.

The retaining of objects as a means of preserving chosen elements of the home also becomes part of wider negotiations with the home's pasts involved in practices of *making* home, as will be explored in Chapter 7. This relationship, as we will see, is complicated by contemporary ideas of value.

Note

1 Here the garden as a site of death and burial reflects complex relationships between animals and humans (see DeMello 2016; Kean 2013; Maddrell and Sidaway 2010; Shuurman and Redmalm 2019). Miss H's cats' remains are left intact out of respect of the assumed strength of *emotional connection* she had with them; the stones mark a territory not to be disturbed. The meaning of Josie's troubled reaction to the thought of giving her own pet cat a memorial stone is unclear. It appears to reverse a normative cultural response to the *absence* of such stones for the recent *human* dead (although see page 193 for an alternative response to an animal burial). This wish *not* to have a stone for their cat does not imply that they have mourned its loss any less, but without a stone they leave open the possibility that future residents might dig up their cat's remains, just as they had dug up other animals in the garden.

References

Armitage, S (2012): *Walking Home: Travels with a Troubadour on the Pennine Way*. London: Faber and Faber.

Bachelard, G (1994 [1958]): *The Poetics of Space*. Boston: Beacon Press.

Billingsley, J, Harte, J and Hoggard, B (2017): *Transactions of the Hidden Charms Conference 2016*. Hebden Bridge: Northern Earth Books.

Blakely, H and Moles, K (2019): 'Everyday Practices of Memory: Authenticity, Value and the Gift'. *Sociological Review* 67.3: 621–634.

Blank, T, and Howard, R (2013): *Tradition in the Twenty-First Century: Locating the Role of the Past in the Present*. Logan: Utah State University Press.

DeMello, M (ed.) (2016): *Mourning Animals: Rituals and Practices Surrounding Animal Death*. East Lansing: Michigan State University Press.

Harries, J (2017): 'A Stone that Feels Right in the Hand: Tactile Memory, the Abduction of Agency and Presence of the Past'. *Journal of Material Culture* 22.1: 110–130.

Hill, J (2007): 'The Story of the Amulet: Locating the Enchantment of Collections'. *Journal of Material Culture* 12.1: 65–87.

Hoggard, B (2019): *Magical House Protection: The Archaeology of Counter-Witchcraft*. New York: Berghahn Books.

Kean, H (2013): 'Human and Animal Space in Historic "Pet" Cemeteries in London, New York and Paris'. In J Johnston and F Probyn-Rapsey (eds.) *Animal Death*. Sydney: Sydney University Press: 21–42.

Lipman, C (2019): 'Living with the Past at Home: The Afterlife of Inherited Domestic Objects'. *Journal of Material Culture* 24.1: 83–100.

Maddrell, A and Sidaway, J (eds.) (2010): *Deathscapes: Spaces for Death, Dying, Mourning and Remembrance*. Farnham: Ashgate.

Massey, D (1995): 'Places and Their Pasts'. *History Workshop Journal* 39: 182–192.

Owens, A, Jeffries, N, Wehner, K and Featherby, R (2010): 'Fragments of the Modern City: Material Culture and the Rhythms of Everyday Life in Victorian London'. *Journal of Victorian Culture* 15.2: 212–225.

Shuurman, N and Redmalm, D (2019): 'Trangressing Boundaries of Grievability: Ambiguous Emotions at Pet Cemeteries'. *Emotion, Space and Society* 31: 32–40.

7   Improving home
The ethics and aesthetics of custodianship

Reburying inherited objects and retaining fixtures were forms of acknowledgement that other people have, and will, experience their own time in the home. These were acts of respect for the home itself and part of a sense of custodial duty to safeguard it for those both before and after. For many participants, any *renovations* made in the course of home-making must also be judged to be an *improvement* or enhancement of the home's physical spaces. Such practices of modification, and the many situated and contingent factors surrounding them, take place in the context of different degrees of consciousness of the broader genealogies and presences of the home's pasts; the practice of home-making, as will be explored here, is more straightforward for some participants than for others.

Sometimes participants described personal reasons for wanting to put their own 'stamp' on a home. Jacob in the southern English town, whose ex-wife previously lived in the house with her second husband, reflected: 'There is an element of overriding the alterations that she – *made*. It's putting our mark – you know, it's a kind of psychological and emotional issue'. Pam in Yorkshire pointed to the track she had built from the only road in the dale. This elegantly snakes down the hillside through the fields to her house at the bottom of a valley, by a beck – quite literally remapping the landscape: 'I've just changed the face of the planet. I've put in a road that you can see from space that wasn't here before, and I've caused that to happen … I've changed the way the world looks, that will still look different after I've gone. And that felt quite a privilege'.

This desire to leave an imprint contrasted with her ecological belief in the need to tread lightly. But she explained the personal reason for her pride. Her mother had been cremated and 'doesn't even have a grave … so she will have left no mark on the earth … When I die no one will remember my mother'. The mark that *she* has left – significant enough to be seen from *space* – became a tribute to her mother. Like the planting of a tree or burying of a pet next to gravestones, this is a different form of private memorial that doesn't need to be witnessed by others – for whom it remains merely a functional track.

Home personified

For others, it was important to emphasise that making improvements to their home was not motivated by a desire to be remembered. Adam in east London pointed out that in 350 years' time 'nobody's going to give a toss, are they? I mean, nobody's even going to think about us, are they? ... It's not that I have to have this legacy. I don't need anybody to remember me when I'm gone'. Martin likewise argued that although he was proud that 'we're actually making the house better, mending it if you like', this did not need to be witnessed:

> And no one else would see me do that ... It's the same way you'd nurture a child in a way ... to come into this world and grow ... I'm not stamping *my* identity on it. I don't feel I'm doing that ... I'm improving the place *itself* ... more than I felt that I was saying, 'This is me'.

'No one *else* would see me do that' – akin to ideas of a quotidian construction of the past that 'does not seek to attract an audience' (Robertson 2016 [2012]: 2). For Martin, the mending of the home is another form of private, visceral encounter with the home, here deemed a 'small child' that needed to be 'nurtured' to 'grow' – an image reflecting a paternal, protective instinct, and one of many ways people personify their relationship to home. One participant compared her home to an *old* person, another perceived the inherited objects and décor as like the 'character' of someone who has 'wrinkles' because they have '*lived*'. Penny described her house as like a 'grandparent', containing 'strength and wisdom'; taking care to replace a window with an original-style one was giving the house a 'present'. Elsewhere, home becomes a series of lovers, or the one true love, but, as described, it is also 'dead' without people in it. Like those who believed the emotional temperature of their home related to the past events it had experienced, others described homes as having 'feelings'. Susan spoke of the 'feel of the house'. Ben's house looked 'lonely and forlorn', Lillian's was 'very sad and neglected and I wanted to make it beautiful', whereas Adam described how 'it has a really *nice* feeling, this house. And this house has never let us down'.

The sense of the home's active, engaged agency – its human-like qualities – offers further opportunities to make connections, reflected for example in the way some saw their characters converging. Gemma, as described, believed her cottage to be quirky and only attractive to eccentric people like herself. Pam in Yorkshire personalised her relationship to her farmhouse, describing an affinity with the fact that it 'sits low in the world. It's a small, low house and that pulled me in. Because I'm a small, low person'. She felt the 'house *chose* me', adding:

> It felt like a home waiting to be created. ... Is it Le Corbusier that said that houses are machines for living in? ... And I think that's bollocks,

basically. Houses are not machines. Houses have personality. They have atmosphere. ... The house seems to have a life as well. It seems to have a presence that needs respect, requires, *demands*, respect. ... [But] it's acquired a *different* personality by what I've done to it.

The house *chose* her, but then allowed her to *change* it. Extending this, participants described getting acquainted with their new home as a form of communication with it. Sandra explained: 'You can decorate to your own style, or you can let the building *tell* you'. Moving to a house after years living in a council flat, Julia didn't rush to impose *her* personality either. She sat and 'waited':

I mean when I moved in here first, I used to sit here. For the first six months. I mean there was very little up on the walls, deliberately. I thought I'd wait. I'm not doing anything until I've got used to the place. And I know what it's like. So I used to sit here. And I used to gaze around.

She kept the home bare of her own clutter to allow the house to be known to her before she personalised it; a desire not just to *fill* the new home but to *fit into* it – a *self* respectfully yielding to an *other* before accommodating itself.

For Jack in Bristol, a new home is like meeting a new person, and part of this process of getting to know it is to understand the history:

To me a house is like a person ... [its history] a bond between me and the house ... It's like when you meet somebody you get to know their *history* ... It builds a picture. Learning about the house's history is like getting to *know* it.

Custodianship

This degree of personhood or agency granted to the home reflected attitudes to it and caused participants to dwell on what made a home distinctive, including how it had a life *of its own* (Yarrow 2017, 2019). As Lillian said: 'I've always felt that they're rather special and that you – can never really *own* a building', and Cathy warned that her house 'can't be held responsible for what happens in it'. This, in turn, increased senses of responsibility to the home, the feeling of being its custodian or caretaker – an ethical stance reinforced through reflections on their *temporary* place in the wider genealogies of residence. For Lillian, as described, she is the home's 'guardian' – 'You're just the custodian of it. I'm passing through and then the next person'. Yolanda said the same: 'I'm just a curator, aren't I? I'm a caretaker of this building. That's all I am'. Cathy described feeling 'privileged' to have a home, but was herself 'insignificant', part of something bigger. Julia dwelt on the responsibility of being a custodian to renovate her 'unloved' home, adding: 'Because it's an old house ... I realised that I felt very much like just a *custodian* you know. I'm just the current custodian … a very small part of it … It is a responsibility [to] take care of the place really … for the future'.

This stance added pressure to make the *right* choices, not just for themselves but for the home and its other residents, as Chris argued:

> The world, life, isn't just about *you*. You're part of a bigger picture. You know, you're here now. … You've taken on a house that other people have lived in the past, you hope to hand onto people to enjoy in the future and you're just here to sort of hand it on in as a good or better a condition than you found it – much better in our case, you know [laughs].

Lillian used the language of sustainability to describe 'our generation's contribution to make it a better home for future generations', and the language of religion to describe her 'missionary' zeal to 'reform the house', to 'make things better … and more enjoyable'.

Custodianship, then, is not just a position but a *practice* imbued with values, to make the home better for the future. Penny, in fact, rejected the idea of 'custodianship' because it suggested to her passivity:

> I'd say we're more than custodians because we have an *active* relationship with the house. If the house were a museum, which it is not, then we would be custodians, and make sure it didn't fall down, and dust it from time to time. We wouldn't be thinking about what *we* might do to it.

Valuing 'the past'

Like attitudes to inherited objects, what might – or should – be *done* depended on the values granted to the past of the home as a physical structure. On a purely practical basis, some participants complained about the 'compromises' they had to make to live in older homes, especially those who had discovered theirs was a 'cash-eating monster' (another metaphor giving home agency) or, like Ben, had bought a 'complete wreck'. Jacob in southern England described how in his last job he had been 'lumbered with historic buildings' that proved a costly burden. The experience had made him 'hesitant about taking on a building of this age, of the structure', because he 'knew enough about them *not* to want to … live in an old house. And the surveyors' report – *silenced* me. *Gobsmacked* me. But we still went ahead'.

Others reflected further on society's 'return' to interest in the old, pointing out estate agents' assumption that old houses have more 'character'. Karen argued that this idea – that 'period is better' – was due to mere 'snobbery', adding: 'You pay a lot more for period, so people must *want* it more'. But for most participants, older buildings were considered better designed and higher quality, with original craftsmanship often pitted against the negative value given to anything 'modern', meaning anything post-1960s. There were exceptions. Ben complained about the original Edwardian fireplaces he inherited being 'cheap as chips ... just old bits of metal, they are not very well made and they don't look very nice either'. Nigel reflected on the assumption that 'traditional houses are seen as being better constructed'. 'It's an old house, it's been here all these years, it must be good quality, it's lasted all this time'. He added: 'The thoughts are that ... they're going to survive a lot longer than modern buildings ... The question is how *much* longer are they going to survive [laughs]?'

But most people venerated the standard of the original craftsmanship, the 'properly made' artisan work, as Karen put it. Alex explained: 'There's something to do with modern buildings and modern furniture just not being up to scratch, being made as cheaply as possible. It's craftsmanship and pride in your work with old buildings'. And Jacob added: 'You're talking about real craftsmanship ... handmade, not fit yourself with a quick sort of screw through. The dovetail joints, the wood is quality'. Chris added: 'You sort of appreciate the craftsmanship that went into making it', pointing to the structure of his early-19th-century house.

This apparent nostalgia for the old, particularly the pre-modern – vaguely the time before mass production – was also connected to an assumed *ethos* which those undertaking hands-on renovation work attempted to emulate. Pam in Yorkshire described the original builders' attention to detail as 'thoughtful and creative'. This also related to the choice of where the house was made to respectfully integrate or 'sit' in the landscape – in a way that 'the land accommodates it'. She argued:

> It is respect for the land; that the people who built these houses understood the land and they didn't try and dominate the way that other societies, cultures, do try and dominate with their buildings ... it's also organic. It's almost as if the land created the house itself, it's just grown up out of it.

Here the quality and practices around historical modes of building stood for the 'understanding of the value of things' as Leah put it. She, again, pointed to her knowledge in the shift in sensibility over the last twenty years to one which didn't rush to demolish, destroy or replace things for the sake of it. This belief in a collective shift to value the past afresh as part of a wider ethic of care, influenced specific responses to the more *recent* décor and features, about which people felt able to be more judgemental, and which, in turn offered them more leeway to make changes. In contrast, the home's earlier or original elements tended to be respected more, leading to feelings of inhibition when wanting to make changes. But the basic equation – the greater the distance from *now*, the greater the value granted to a home's materiality – was complicated by participants' different aesthetic preferences. Some cherry-picked a favourite era, leading them, as described, to skip over those deemed less engaging. The Georgian era was the most popular. Abigail lauded its design aesthetic for getting it right – the quality of light, the elegance – something which was 'lost later on when the styles changed'. Some participants were also impressed by the Victorian style, but, as discussed, not all. Ben complained: 'Some of these late Victorian, Edwardian houses have – very *fussy* – sort of cornicing. This didn't have it. Which is great … I prefer plainer'.

The home's original 'integrity'

Whichever ways its 'pasts' were differentiated, defined or valued aesthetically, many participants felt it their custodial duty to make choices, to decide what to embrace and emulate or what to ignore, discard or even condemn of the home's material inheritance. But how do people act as custodians of homes which need to be *lived in*? The responses, explored below, fell into two broad camps: a belief in the importance, above all, of respecting the home's *original* aspects as authentic, and a belief that a home should be left to *evolve* over time. These positions informed attitudes to home-making and restoration, although in almost all cases compromises were made which required reconciling.

For some, the home's 'identity' directly related to how it was *built*, its original fixtures, fittings and structure. Participants used words such as 'purpose', 'intention', 'integrity' and 'feel' (reiterating the language commonly used by building designers and estate agents) to encapsulate a belief in the home's identity as an intact whole. This led to a need to protect or restore original features. Brenda in east London expressed this as an attitude of restraint, a holding back from making too many changes:

> I don't like doing anything too *drastic* to an old house unless it's absolutely necessary. Unless there's a *very good reason* to do it. I have this idea about the *integrity* of the building. I know it's hifalutin for a fairly low end Edwardian terrace but – I think houses and buildings are built as a *whole object* in a sense. And if you're not careful – it starts not to work properly.

Yolanda likewise described a responsibility to the building's original intention, which she defined as 'how it's meant to be. Its purpose', whilst Lillian reflected on a desire to 'introduce the *feel* of it … like by putting panels in … the feel of the original house'. Harvey and Rochelle's restoration project focused on replicating period fittings to create the way in which their house was 'designed to look like originally'. By researching design details, they could 'conserve things from the past'. Pointing out how he had 'exactly matched' the originals when restoring fireplaces, coving and doors, Adam commented: 'There needs to be some sort of integrity in the house, doesn't there?' And Jane and Nigel kept original features to

> restore the *integrity* of the property as it was built. It was built in a particular way as it was originally put together, and it's had a few changes

over the years, and it's trying to go back to that original integrity of the place, I think. It was designed and built with a particular standard, a particular function when it was put up. And I think the idea is to try to *keep it* as much as possible to that original design, the original build. Even though it was a hundred years ago.

(Jane)

'In keeping' and 'fitting': pastiche and reproduction

This belief led to judgements about the way homes had been modernised – a particular focus of disapproval being the replacement of original windows with plastic ones. As described, Pam would never put in a fitted kitchen, which would not be in keeping with her farmhouse. Where original fixtures had already been removed, for some these needed to be restored using, as Cathy said, 'absolutely right replicas'. Chris and Josie had replaced a fireplace with one from the original era, and Anna had also replicated as closely as possible bay windows in her living room, despite the difficulties she encountered finding the exact colour matches for stained glass embellishments: 'These are just so beautiful and so important. It is part of the house'. But, as Josie had emphasised, it was also crucial to be clear what was truly original to a home and what was not. Anna made the same point: 'I don't think you want to … try to *fool* people that they are original'.

For many, it was more important to honour the original *look* of the home. Lillian, for example, argued for the need to 'introduce the feel of it … I wanted it to be done sensitively and harmoniously … You can't be too precious, you know. Just give a *feel* of it'. Jack in Bristol was scathing about the inappropriateness of putting modern furniture in an old house, which wouldn't 'suit the house … You have to be in sympathy with a house, don't you?' But, again, this didn't require choosing original furniture; it was more important to get the *feel* right, 'the feeling of a bit of age and a bit of history to it'. Adam also believed that honouring the home's 'integrity' meant matching original features with those 'in keeping' – but he felt it important to find an 'exact match' *where possible*. Nigel also described how restoration meant 'doing things in keeping *as much as possible*'. Harvey and Rochelle salvaged original features from local builders they had befriended who were gutting local houses to turn them into hotels. The fittings, he said, had to 'match' and *look* like a 'precise copy'. And for Dan, living on a main road in east London, the issue was also one of choosing a look that was 'in keeping'. He said:

I've put sash windows *back in* … The windows were terrible. It's almost like the house *deserved* it … to look in keeping with the house and the street. That's how it was built, and that's how it should look.

For Emily and Derek, however, their furniture had come first. They chose their unusual arts and crafts house in order to 'fit' their furniture: 'You buy a house to suit your furniture. At least we did … All this furniture has a *history*, every piece of it ... This is my great-grandmother's arm chair'. As well as physically fitting the spaces, their furniture also had to fit *aesthetically* – much of it was wooden and 'of the style' of the house; what they *brought* to the house needed to *match* the house. But there was also a sense of the house itself resisting change. Its quirky structure included curved ceilings that couldn't be replaced with plasterboard and wonky, crooked walls which defied modern *fittings*. The architects had also 'fitted out' integrated bookshelves, nooks and crannies and even a bedframe – dictating in advance how bodies and their belongings should fit the space. Attempts to make changes were challenged by this original structure; the house fought back against modernisation:

DEREK: A lot of people don't like the fact that there aren't any *straight lines*.
EMILY: Some *workmen* won't work on this house because it doesn't have straight lines.
DEREK: You could modernise it, and also at the same time, *destroy* it.

For most participants, then, reproduction features were acceptable if they were 'in keeping' with the original or sensitively 'fitted' the original space, as long as there was no pretence that they were original. Jacob in the South of England, for example, was scathing about 'repro Victorian' bathrooms with roll-top baths – particularly ill-fitting he felt in their Georgian house. He hated the idea of 'pastiche'; it was more important to have 'confidence in our own taste'. Susan also put the aesthetics of her Edwardian house before her own, stopping herself from replacing a 50s fireplace with a Victorian one she had preferred: 'We bought something that's not as decorative ... It felt like it needed something plainer than the ones that I would have liked'. But this was still 'pastiche', she conceded, and therefore not completely respectful, although it fitted the *look* of the room better:

> I don't think it's respectful in the sense, I don't think it's *correct*. I mean, if you know about it, someone who knew about it could come in and go, 'Well, actually, what you've chosen is a Georgian pastiche of a – '. It's not correct in that sense. But it was our attempt to choose something that might have looked right, *there*.

Those whose historical knowledge had evolved since living in a home often changed their attitude to renovation, regretting what they saw as past mistakes. Leah's research came too late for the 'right' door handles:

> It kind of grates on me ... all the door handles are wrong ... shouldn't be like that if it were a proper Victorian – you'd get it right, you'd put the round ones on. That annoys me. But you wouldn't go round and change every single one, because there's probably about eighty of them [laughter] ... I know it's not right. When I've read all these books, I'm like, 'Oh no, it's not quite right'. But never mind.

Knowing a historically 'correct' style can also become an inhibiting factor in making home, creating a tension between what is *desired* for a space, and how it *should* be done. Yolanda regretted the choices she had made before completing a degree in design history as a mature student; her academic study of history shifted her attitude and values. She now understood the difference between restoration (a 'dirty word') and conservation, and in turn the worst sin of all – to create 'pastiche'. She had copied a light feature from architect Charles Rennie Mackintosh's house, taking pleasure in it and the way people had reacted to it: 'Lots of people say, "Wow, I've never seen lights like that anywhere"'. But she now condemned the lights for being a bad copy of an original and for not being of the right era:

> That's completely wrong for the house. ... These are post-modern items in my house. ... I don't like the idea of post-modernism. ... The fact that I've stolen something from a previous era, and not only stolen the idea but changed the idea. ... [The original architect] would hate it. ... I would never do this again. But it would mean ... don't tell my husband – getting rid of the lights, and they cost a great deal of money. ... I think what rankles is that I think I got it *wrong*. I got it *wrong*. ... Am I not respecting the house by using these lights?

Home as palimpsest: not a museum

Yolanda's regret about her 'post-modern' lights is, however, tempered by her understanding that some of the earlier fixtures may themselves have *predated* the house. The lights might have been a copy from a design a few decades before her house was built, but, she mused: 'most houses have some objects in them which come from a previous era ... prior to the date of their house'. Brenda, also a stickler for maintaining a house's 'integrity', made the same point about a fireplace: 'I would say that was a Victorian fireplace, and this house is Edwardian. So the fireplace looks like it's older than the house. It's possible that whoever built it put in old fireplaces from the back of the workshop, kind of thing [laughs]'.

People's interest in tracing clues about previous changes to their homes further acknowledged the fact that homes are modified over time, lending itself to an alternative view of the 'identity' of homes. Lillian described how she was 'interested in how things are shaped and how things evolve', whilst Anna stated: 'It's about – not just a house, but how it's changed. And who's been there'. And Jack in Bristol also expressed an interest in who built the house originally, but added that he'd 'love to know how the house has evolved'.

This interest is reflected in many participants' attitude towards making changes to their own homes, becoming an ethical as much as a pragmatic position. Dan explained: 'I'm not in favour of keeping things as they were', whereas Lillian, perhaps a little more defensively, added: 'most houses evolve anyway, don't' they? You might have a modern door and Georgian windows'.

Thus measuring a home's identity in relation to its original form is deemed too reductive, ignoring the ways in which different residents have placed their physical mark on it over time; if what is inherited has pastiche elements or mixes styles from different eras – this constitutes a true reflection of the home's material biography.

Some, like Samantha, in her flat full of salvaged objects, *revelled* in a sense of a home's layering of different eras. She said:

> I like the mix of different periods, the sense of the house as a kind of *palimpsest*, that it has lots of different layers, that it's not – you know, some people, when they restore a place, want it to be *perfectly* what it was – at a single point, whereas I like the feeling of this having been *lived* in for generations and each generation changes it. But there's still traces of it all.

Rita made a similar point: 'I just know from the very beginning, people were changing that house. And who am I to say – "the original was really like this"? You would have to knock down half of it, you know. I mean, it's silly. A house, you know, reflects the changes in the fashions of people that live in it, and the needs, the changing needs'.

Some felt relief rather than regret that there were so few period features remaining in their homes. Karen commented that features such as ceiling roses and cornicing collected dust and were hard to clean. Alice also said she was 'pleased' her house's features had been removed: 'We're both practical people and we wanted it as easy as possible'. Brian and Megan felt their home was 'too new' to be historically interesting, preferring to focus on the history of the previous house on the grounds. Older houses 'demand so much of you', Megan commented: 'I want somewhere comfortable to live, that's easy to look after. I can get my history from somewhere else'.

This progressive view of the home's past offered some people leeway to feel justified in making alterations. They argued that heritage of a *home* should not be confused with the more static, and definitely *not* lived-in replication of past domestic interiors found in *museums*. Rochelle argued that she didn't 'want to make it a museum but a home that we could live in'. Susan parodied the 'living' museum trend: 'I don't feel a need to … to go *back*, not like those Spitalfields houses where people don't have electricity [laughs] and stuff like that'. Yolanda added that even historic house museums often choose which era to 'restore it to or conserve it to', which may not be the original era. And as Penny had noted, being a custodian implied a passive role of looking after and preserving a place whilst, in any case, it was impossible to restore a house to its original:

> If you did make it happen, you wouldn't have a house anymore you'd have a museum – because you wouldn't usually want to live in it. Unless you are running a little experiment for school kids to see what it was like living in 1792 with pitchers of water and taking the chamber pots out with the night soil!

Accepting compromise: selective decision making

Others argued that in order to live in an old home, there would always have to be compromises. The emphasis on making choices 'in fitting', 'in keeping', or in the 'spirit' of the original can become part of the justification required to grant leeway during renovation. Nigel described how he was more interested in 'having a *comfortable* home than in reinstating everything', admitting this meant paying 'lip service' to the home's integrity:

> Things do *move on*. But I think you want to do things *in keeping* as much as possible, while taking account of the need to move on a bit in terms of better practicality. ... To some extent it's paying lip service, because this house has been extensively changed, in common with most other houses. Unless they are museum pieces. ... That's what you have to do with a house.

Some participants, such as Derek, compartmentalised home-making and an appreciation of the home's 'integrity' – 'Not to stay as it is, but its *integrity*, its *spirit*' – the two not necessarily incompatible. In any case, as Abigail pointed out, the original Georgian owners would have liked to think they were being 'modern' on their own terms; she was continuing *their* modernising ethos. Cathy described her attitude to compromise:

> It's a bit of a mix and match. You have to get on with life, in a house. But I'm keen it retains some Victorian integrity. I think I'm a bit of a balance between the two positions. I do think you can mix your eras, you just have to get the balance right. I do mix and match ... I do *disrupt*.

Pam also described the importance of retaining the original character, identity and integrity, having researched books on old houses and indigenous architecture to 'gain some understanding of what it would have been like'. She wanted to maintain the original layout which 'worked beautifully', and particularly aspects that would have been familiar to earlier residents. These included the niches in the stonework that her stonemason had decided to fill in. She removed the stones herself to preserve the feature:

> I suspect they were places where they put candles or whatever ... They wouldn't have had many possessions, they wouldn't have had much furniture, so these places were important. ... It was critical ... that the main house had to retain as much of its original character and shape as possible. I didn't want to overtly modernise it.

But she also added partition walls and compromised over modern technologies – running water, central heating, 'the bathroom'. The renovation of the farmhouse had been a deeply personal process as her partner had died before she started the work – it had originally been a joint project. So she also added chrome to the banisters, radiators and stoves in her memory: 'She loved these materials. It's me saying, "Okay, I've got to have chrome"'.

The decisions Pam made in renovating her farmhouse had to accommodate different needs, values, beliefs, understandings, memories and relationships. Other participants found their own compromises, often down to the details, piecemeal, fixture by fixture, room by room. But patterns emerged suggesting particular values granted to some rooms and features more than others. For most, windows should not be modernised – nothing else but the original would do. But elsewhere 'concessions' to modern comforts were allowable. Practical or financial considerations often outweighed all else. Lillian added a balustrade at the top of her home which was fifty years later than the house's original date because she could not afford to use the original design. She also 'straightened' a floor, even though she didn't believe this was the right thing to do ('this is not a straight house') because it sloped so much the television 'travelled'.

Emily and Derek were critical about the previous residents' changes to the house, which disrupted the house's originality. But they were happy to maintain these, admitting they were practical. The original brick fireplace had been painted white but, Emily said: 'I'm not taking it off because it does actually make the room lighter'. And they did not wish to reinstate the removed fireplaces in the upstairs rooms because that would mean 'you wouldn't have very much room for storage'. They had replaced the 'nasty 50s mess' of a kitchen, but did not reinstate the original sink, even though this was available – 'down the garden full of plants'.

Colour schemes were also the subject of compromise, sometimes surprisingly given the amount of research undertaken to replicate the original aesthetic. Rochelle had completed studies in archaeology and conservation, absorbing a desire to return the house to its original and wanting to get the restoration right, including choosing 'wallpaper and colours that would be in keeping with the age of the house'. She explained:

> I had a straight interest in conserving things from the past, which in my book means keeping as much of the original as possible. And if you can't because the door's been cut down, then at least try to keep the original door somehow, and use it somehow. And if you can't have the original in the front of the house then you make a precise copy. So that, you know, is fully in keeping with what I would have learnt about conservation from museums. But we … wanted to make a home that we could live in. … And it's quite true that this apricot colour is not a Georgian colour. It was chosen because of the colour of the sofa.

Despite the research, it was more important to choose a colour to match an existing modern sofa than to adhere to original colours. There was also the issue of which rooms could be allowed to escape the return to the original; most people didn't give kitchens and bathrooms a second thought. Indeed Harvey installed an 'ultra-modern' bath: 'That's the sort of thing you don't compromise on. You don't go backwards on those'. Rochelle was a little defensive about this, wanting to reflect on the *overall* impression of the restoration:

> I think this house can say to itself – it is a Georgian house. *Still*. Yeah ... Most of the things we've brought into it [pause] – *fit* the Georgian character. And the things that don't, like the kitchen and the bathroom, it's *easy to understand* why they don't.

These rooms did not count, they are exceptions governed by different rules. In Rochelle and Harvey's house, two further spaces were allowed into the same category. Firstly, the basement dining room was relegated as the space for modern furniture, accommodating a family inheritance which had an emotional value:

ROCHELLE: Down in the dining room we've got furniture from the 60s which I inherited from my parents. That's the one room where we have got modern furniture. It didn't feel as old as the rest of the house ... Harvey actually had some quite strong views about not having things that were not *old* in the upper part of the house.

HARVEY: Well you didn't want to get rid of it.

ROCHELLE: Yes ... I suppose that was the compromise that we came to. That we would have modern furniture in one room.

The basement room was not on public display, and became a kind of spare room for personal objects that didn't otherwise fit their project. Not feeling 'as old' as the rest of the house, it was also at the bottom of a vertical hierarchy of historical value. The second devalued space was, as described, their garden – considered a blank slate which thus offered space to explore personal passions. Harvey had created a Japanese Zen garden, replete with raked gravel, rocks, small trees, and a framed view over 'borrowed' scenery – in their case, of a council estate at the back. The contrast in style and aesthetic, walking from house to garden, was striking: 'The point about Zen gardens is – that you can fit them into an extremely small space. They can be very meaningful ... Whether it's really in keeping with the Georgian house or not, I doubt [laughter]'. His personal enthusiasm for Buddhism and nostalgia for memories of Japan trumped any slavish adherence to their Georgian project, although, as he admitted: 'I like to pretend it is in [Japan]. ... Now the borrowed scenery here *isn't all that great*'.

Feeling guilty

Despite these compromises, some participants expressed discomfort. Martin felt that they had been 'selfish' for aesthetic reasons, noting that they 'couldn't afford the time or the effort or the money' to return the house to its original style, but that probably more importantly, from our own selfish point of view ... I can't live with twee little bits that were prevalent in 1839 and 1840 ... And since it had already been ruined ... we weren't fighting against anything, if you see what I mean.

The dramatic image of 'fighting' the home's original identity (or avoiding doing so because original features had been removed) is a further way in which the home is granted agency; it is the home itself that people imaginatively tussle with over the need for compromise. But elsewhere, the home becomes something more vulnerable. In the southern England town, Jacob and Penny's main project was to remove a huge chimney which the Victorians had added in a prominent, central place straight through their many-storied house. The removal of the chimney and fireplaces required cutting a hole through the middle of the house from 'the basement right up to the top ... a hole the size of that sofa, at every level', Penny explained. She was concerned about their responsibility to the house, wanting to protect it:

PENNY: When I realised I was a bit nervous about what it was going to mean. Because it felt to me the night before it started as though the house was going to have an operation [pause]. So I did feel a bit nervous about it.
CL: That was ripping something out?
PENNY: Yes, yes. Just exactly that ... The sense that – we had some control over what was going on, but actually – not that much. And just this kind of feeling that you hoped it would be alright ... When this [chimney] has been dragged out [quieter voice] – how will it be? How will the house be? How will it feel? What will it be like for *us*?

There is slippage between the feelings of the house and the feelings of the occupants – 'how will *it* feel? What will it be like for *us*?' But Jacob emphasised that the removal was justified because it was not *original* – it was an error by a later generation which served a purpose for them but was no longer useful. Removing the chimney honoured the original alignment of the space: 'I suppose I got the planners to accept that by saying this is a Victorian non-structural insertion. ... It was a minor awkwardness'.

For some, like Adam in east London, the different values given to different eras also justified decisions. He said: 'If it was a Georgian house, probably I'd feel more likely to preserve it ... We might have been more careful with it'. He also had no qualms about demolishing an old coach house in the garden because it had been 'really kicked about ... it wasn't nice enough'. But when it came to removing the original pews in a ruined old chapel he had purchased with his siblings, his response was radically different. This gave him a 'strange' feeling of having 'transgressed'. The unease was in part visceral – the pews proved solidly nailed to the ground and required coaxing out: 'You had to rock them and rock them, and keep rocking them until the nails broke. And then you could get them out. It was a really strange feeling, taking them out. They'd been in there since 1857'. The effort required in removing the pews suggested to him the value of the original craftsman's work; they had been 'beautifully done'. This reinforced his feelings of unease: 'I had a funny feeling when I did it, like it was a crime, somehow'.

Elsewhere, Chris and Josie seemed to sign up wholesale to the palimpsest view, considering this to be the 'natural' state:

CHRIS: Some people want to strip it back to how it was when it was built. Other people won't like appreciate all the changes that have been done over the years. And I think we're probably quite happy to live with the changes because it's just a *natural progression*.
JOSIE: It's part of the history of the house.

They mocked popular takes on 'restoration', pointing to an example of a recent television programme:

CHRIS: Some of these house restoration programmes on television … You know, they'd tried to restore whatever aged house it was. But then you walked into the bathroom. It was *space age*. And you think, 'Well' –
JOSIE: Yes, with flashy light and things.
CHRIS: You know, 'where are your principles here?' You know [laughs]. It's unrealistic, isn't it?

But their tone became more defensive when recalling the Victorian green-house in the back garden, removed to accommodate a parking space:

CHRIS: Big Victorian greenhouse. It was lovely.
JOSIE: Yeah … I have to say it would have been very difficult to preserve anyway.
CHRIS: But that was a hard decision to make.
JOSIE: That was hard … Street parking adds value to the house.
CHRIS: It would have been impossible to take it down and rebuild it. It just wasn't feasible … But also street parking is so difficult.

The 'hard decision' is justified in a number of ways – adding value to the house, probably not saveable, inability to rebuild, a practical necessity. Nonetheless, there was regret requiring a touch of justification, as there also was for a thrown-away meat safe, as Josie added: 'But it fell apart. It was completely worm-ridden, wasn't it? … We feel quite *guilty*'.

Fireplaces

For some participants, finding evidence of decisions made by *previous* residents, which revealed their *own* relationships to inherited or original features, offered not only insights but also further justification for change – suggesting as it did the concessions *they* made to the fashion and mores of their own era. Chris and Josie, for example, pointed out that the Victorians had replaced some of the banisters of the main staircase in their Georgian house. They had placed mahogany downstairs, but retained the original one upstairs, which was 'just pine that's stained'. The mahogany had stopped short at the point that visitors could no longer see it when entering the house. There was no need to replace the top section as these wouldn't have been on *show*. Chris explained:

> This is the bit the visitors would see when they arrive. Whereas the original Georgian bit, you know – obviously they didn't think visitors could see that so they didn't bother to replace it all the way up. ... It's sort of typical of their attitude at the time ... The best quality materials were used downstairs with lower quality upstairs. And even down to the window catches downstairs are glass and upstairs they're iron.

This well-known Victorian mind-set – where the better quality materials were reserved for the areas on public display, including front rooms being the 'grandest ... the impressive ones for when guests come round' (Leah) – helped some participants themselves differentiate between the more public and private spaces of their homes. They reinforced this distinction in relation to the choices *they* made (Goffman 1990 [1959]) – explaining why some rooms became a greater focus for renovation efforts, allowing other rooms more leeway to be 'modern'.

Fireplaces and surrounds were the focus of much discussion.[1] Some participants, as described, felt it important to replace these with original-era ones where these had been changed. They were seen as particularly resonant anchoring focal points for imagining previous lives, as Susan reflected: 'When you look at that fireplace, you can think to yourself – if you're being fanciful – you know, someone must have had to get heat, to keep the house warm someone would have been there tending that'.

But apart from their symbolic value in relation to their functional and social role in homes past, fireplace surrounds are unusual in being quasi-permanent fixtures which can be moved without risking structural damage to a home. Often, they were found to have travelled *within* the home, moving from one room to another, reflecting the different values granted both to the fireplace surrounds themselves and the different rooms in the home. A particular strategy in balancing a respect for the home's identity with personal taste preferences was to remove unwanted fireplaces out of sight to *lesser-used* rooms.

Cathy had described how 'we're very keen on preserving the originals. When we had the rooms knocked through ... it cost us a *fortune* to replace the cornicing along both sides of the middle. ... That is all reproduced'. But she also loved her 1930s art deco gas fireplace she had taken with her from home to home, stating:

> I love that fire. When the sun comes through it and bounces off it you get rainbows all around the room. ... I took it with me, because I loved it. It's a faff taking it with you because it weighs a ton [laughs]. I love it.

In order to accommodate it in the living room, she had removed the original Victorian fireplace to the upstairs master bedroom, where, she argued, she didn't have to look at it as much: 'You spend a lot of time up there but the lights are off'. Like the found objects which are kept out of sight but *within* the home, Cathy is keen to point out that she had done the same:

> I *preserved* it. I felt duty bound to keep it ... because we've been so judge-mental of people who had ripped out Victorian fireplaces and trashed them. I thought, 'Oh god, am I going to do the same?' ... So I couldn't take out the Victorian fireplace because it was part of the house. Having said that, I think it might have been earlier, I think it might be Georgian. I think it might have been put in anyway, so I'm not sure it was the ori-ginal fireplace. That's a bit of a moot point, isn't it [laughs]. ... But it did come with the house, it definitely came with the house, so there you go. ... It's in the bedroom upstairs. It's fine up there.

Again, there is justification for removing the fireplace from its original space – as something which is a *part* of the house, it is still *within* the house (added to which it might *not* have been original anyway).

Jane and Nigel reflected further on how fireplaces are removed to bedrooms in order not to offend the eye. But when imagining residents' choices during the 1950s, they again assumed this was an act of expedience rather than one of respect for the house's history. Their home contained four fireplaces, two upstairs and two downstairs. A resident during the 1950s had replaced those in the living rooms with 1950s-style ones – which Jane and Nigel replaced back with reproduction Edwardian-style ones. But upstairs the 1950s owners had only bothered to replace the front bedroom fireplace with a 1950s one. Thus the only original Edwardian fireplace was in the back bedroom, which they felt was probably the master bedroom then. Nigel reflected: 'Why did they not bother with that one? People didn't come into your bedroom ... So it didn't matter if you had something old-fashioned in your bedroom. We've done the opposite. We've gone backwards, whereas they've modernised, haven't they?' Jane reflected that they had, in a parallel move, decided *not* to bother replacing the 1950s fireplace in *their* bedroom, now at the front of the house: 'It is pretty ugly [but] it just doesn't seem necessary [to replace]. It's ugly but it serves a purpose ... It's pretty unobtrusive. We've got used to it'.

Elsewhere, Ben also reflected on his attitude to the fireplace in the front living room – a room used by his children – compared to that in the back living room, where he liked to relax with his wife: 'I don't think I could have lived with that, had it been in *there* [the back room]. We would have had to change this. ... The kids don't care. And I'm not spending the money, you know'. But he speculated that the fireplace surround was probably original, suggesting that the residents during the 1950s couldn't be bothered to move it either: 'Down the eras, whoever's come in here has decided *that* one stays ... Pure chance'.

Assumptions about previous changing attitudes to the inherited past, linked to changing aesthetic fashions, suggested insights into past home-making choices. But these were also contingent upon the particular uses of rooms as part of the embodied practices of living in a home. In Ben's case, the front room had been given to his children to use and this meant the fireplace had been left alone; previous residents must have had their own reasons – perhaps similar. In turn, speculation about who had removed or kept fireplaces and why also reflected participants' own ways of negotiating the home's pasts in relation to their own uses of home.

Value, taste, aesthetics, era

If renovation choices reflected the way particular objects were encountered during the process of making home, this explains why not all those things deemed out of keeping or ugly were removed. But a further factor was participants' awareness of how far their decisions were based on their *own taste* – how far this dictated the type of compromises made during renovation in balancing a respect for the home's material inheritance and the desire to feel at home. They acknowledged that their home-making process was about obeying the cultural mores of the day, just as it had been for past residents, a consciousness of being a part of a *moment* in valuing the past in particular ways. Nigel reflected:

> Why do I think I have to put original features back? Because there was a trend wasn't there to put like panels over the doors. There are places where we've taken those off. You take off the panel and there's a lovely fireplace ... I think it's our generation, we were very into living in Victorian houses. ... It's very weird putting things *back* – old-fashioned things back. Why aren't we putting modern things in? Convention, isn't it really?

This, he suggested, was more about the dictates of style than about history – a similar point made by Laura, who, defending her love of retro, argued this was 'more of an aesthetic need, because it just looks nicer', hinting at a sense of freedom to cherry-pick from a variety of aesthetic choices offered by different past eras, stripped of their historical contexts. But elsewhere, as described, participants had marked preferences for the aesthetics of particular eras with some of these, as Julia argued, being 'better than others' (as discussed, within this cohort, Georgian 'clean lines' were often pitted against 'twee' Victoriana). Against the values granted to particular past eras was the greater evil of 'modern' architecture. William, for example, described the effect his council estate flat had on him: 'When I first saw it I thought, *"how ugly"*. This *box* – not even a roof on it … nothing prettying. This is what it is – a council *box*. This is like coming into an institution'.

As participants reflected, their aesthetic values dictated their attitudes towards the inherited past, beyond any more generalised concern to respect the home's history and to be emotionally and imaginatively impacted by it. This made the process of custodianship of the home more complicated. Making and justifying changes as 'improvements', as part of an ethos of taking care of the home for future residents, needed to be qualified, just as people were also conscious that they couldn't control the choices made by *future* residents (although some tried to do so). As Dan said: 'Somebody else will have it and say, 'That's a disgusting colour', or 'Who's done this extension? [Laughs]'. Ben also fantasised about being judged in the future, just as he judged the previous 'odd' owners' 'shoddy' work on the house. He mocked his own contribution:

We are – the present incumbents, and in a couple of years' time we will move and somebody else will come and say, 'oh my God, you'll never guess what, that's *terrible*. We'll have to get rid of that. We can't live with that fireplace, and that's got to go' … And in 10 years' time they may be sitting down speaking to an academic and going, 'God you'll never guess, we've moved in here – the owners were very, very odd'.

In turn, this consciousness that tastes constantly change also reflected dilemmas about senses of obligation to keep those larger objects that couldn't so easily be squirrelled away in rituals of containment. Cathy in east London reflected on a particular imagined scenario to ponder how she might balance her aesthetic response with an ethical desire to respect, retain and protect the home's past. Photographs of her home's interior from the 1950s showed that the famous artist had papered the walls with wallpaper in his very distinctive style. She was relieved it had gone by the time she bought the house, but knew she would have felt hidebound to keep it if it was still there:

When we look back at this house in the 1950s, you'll see [his] photos are all over the walls of the inside of this house – that *heavy* black and white, typical stuff. And I couldn't have lived with that. It was a good job it was gone. Because you couldn't have *taken it out* … but you couldn't have *lived* with it [laughs]. That would have been *awful* [laughs]. I'm not quite sure what you'd do!

Imposing 'values'

These dilemmas, along with the awareness of the complexity and contingency of home-making choices in relation to respecting the home's 'identity' or 'integrity', is brought into sharp relief in relation to complaints about local authority heritage officials. If participants grappled with making decisions which respected their home's history whilst making home for *themselves*, they felt at the least that they had a *greater* understanding of and sensitivity towards their home's history and character than the heritage 'experts' dictating from outside what they could or couldn't do to their homes. Here, like the decision to disallow Bella and Sylvia to undertake an archaeological dig, the imposition of official historical 'values' was deemed random and shallow – albeit also inconvenient.

Martin and Carol, for example, whose house was Grade II listed, had been told to replicate the 'horrible' windows that had been in place when it was listed in the 1970s. The windows had since been replaced and the couple had refused to make the changes, in breach of the regulations. Carol explained:

> Prior to us moving in somebody in between times had put those windows in to try and sort of improve the aesthetic. ... I have to give control over what I can do to my own home to somebody else who, you know, really actually doesn't *care* ... and has *no feel for the history of the place* or anything. We appreciate the value ... from a historical standpoint, of what the house has got. And we would only want to improve it in keeping with, you know, how it should look and feel from when it was built.

Martin added that people had had to fight the local authority to save the historic square from demolition some years before. The same council was now 'saying we can't do anything to it at all to improve it. Well [laughs]. How crazy is that?'

Susan also reflected on the irony of the freeholder estate's attitude in forbidding planned changes to a house that it originally condemned for not being 'in keeping'. Now the unusual design of the house was seen as in need of preservation, although in any case the estate only cared what the house looked like 'from the outside'. Elsewhere, Gemma's listings application was rejected because the cottage was not original enough, 'not typical Victorian' and there was 'too much added on'. Archivists Chris and Josie also described their lack of 'confidence' in the system after failing to get their home listed too, having been told, 'Your house has had too many alterations'. Bella and Sylvia expressed the same frustration, arguing that as homes evolve they should be shown how this has happened, rather than imposing a random cut-off point after which nothing else could ever be done:

SYLVIA: It's an argument against quite such a severe, 'you mustn't do anything with a listed house'. Because *if* you'd said that in 1200 none of the rest of the house would be here.

BELLA: My feeling is that actually houses evolve and this house is a very good example of a house that evolves. Suddenly they say in 19 – whatever it was: 'no house should ever be added to' – is actually I think wrong because it stops the evolution of houses ... To say this should stay as a snapshot in time, prevents evolution.

Others complained that being listed hadn't made any difference: their houses or flats were still being neglected. Sandra complained that although she'd 'been told that it was [borough] listed, it doesn't seem to change *anything* about what they do'. And Elaine also pointed out that her council flat was part of a Grade II Edwardian mansion block that had been so severely neglected that the last surveyor she had spoken to had 'got scared by the sheer level of work'.

   These brushes with heritage 'authority' raised further questions about who should decide what has 'historical' value and how this should be reflected in practice. Despite their acknowledgement of their different beliefs about and interpretations of the value of the pasts of their homes, participants considered themselves to have *more* historical knowledge, understanding and respect for the history of their homes as current custodians than these heritage 'experts', believing that rigid notions of what gives a home historical value should not be imposed from above.

Public and private histories of home

These battles with planning experts reinforced some participants' belief that their own knowledge, experience and connection with their homes as sites of heritage did *not* hold any value, and that, specifically, the pasts of their homes *they* wished to respect had little in common with 'official' heritage values. For some, this was reflected in a different relationship to 'official' forms of heritage because the line between private and public heritage became blurred, making the project of *domestic* custodianship more complicated. Bella and Sylvia, for example, opened their manor house for occasional public tours, marking it out as a hybrid of home, house museum and museum-as-home. They recalled that initially the family questioned why the 'public' might be interested in visiting their home *as if* it were a museum:

BELLA: [People] asked: 'can we see round the house?'
SYLVIA: My husband had lived all his life here, and so had all of them. And they all said, 'Well, it's just our *home*'. You know. 'What's there to look at?'
BELLA: It was just an everyday – where you live, you know.

But they realised that it was the fact that it was also an ordinary family home, albeit a grand one with a long history, which made people interested:

BELLA: They like something personal ... People like coming round here because it's a family house

SYLVIA: You go to [a local historic house]. It's got absolutely nothing about it at all – you know. It's just a collection of reception rooms.

Beyond fulfilling conventional criteria in being a 'historic house', a further way some participants described their homes as potential sites of 'public' heritage was to mark them out as places where former public figures had *lived*. Again the boundary between public and private was blurred as people responded differently to how to deal with such knowledge – some wanted to display public memorials, in the form of plaques on the outside of their homes, whilst others did not.

Adam in east London, as described, fantasised about putting up a plaque to tell people about the famous music hall star who used to live in his house – if he could be sure the story was true. He wanted to show off his home as distinctive and charismatic. But Cathy was less comfortable, ignoring suggestions that she should apply for a blue plaque to commemorate the lives of the photographer and painter who had lived in her home. She didn't want to 'draw attention' to the house, worried that this might suggest her home was better than others in the area, or place too much focus on a few individuals rather than the 'lots of people' who had lived locally. She said:

It's *fantastic* that history. Several people have said I should try to get a blue plaque outside. I'm not sure I *want* that because that would draw attention to it. I just feel it should *fit in* rather than have a blue plaque or whatever. It's part of the *area* ... I didn't want it to have ideas above its station. It's just one of many houses [laughs] that have lots of people living in.

Julia also expressed ambivalence about requesting a memorial plaque to commemorate Keats' stay in her house – like Cathy, she was concerned it sounded pretentious, giving her a form of vicarious, reflected glory earned merely by the chance act of buying the house someone famous had happened to live in briefly. But eventually she succumbed to pressure and contacted a heritage organisation to seek approval for a plaque, naively assuming this would be a formality. But her request was rejected. She recalled:

I got an incredibly derisory response, saying, 'We have a Keats' house in Hampstead'. Well, he hardly lived there any longer than he lived here ... I mean, a bit longer but, you know? So anyhow, I just sort of left it there.

She mocked their dismissiveness. Keats, like many people, had lived in more than one home, and yet there could only be *one* house museum related to his life:

'We have the franchise on Keats. And we have our Keats' house' ... I mean he must have lived somewhere in – he probably lived in Southwark as well because he was training at Guy's [Hospital] to be a doctor. ... He grew up near Moorgate, that's where he was born.

So she decided instead to create her *own* memorial plaque *inside* the home. On an oblong piece of metal she etched the words: 'The poet John Keats lived here in 1820' underneath which was a short extract from a poem: '... in spite of all/ Some shape of beauty moves away the pall/ From our dark spirits'. This reflected Keats' own 'dark spirits' in the house ('pall' meaning 'gloom'), but the quote expressed a reassuring sentiment about the uplifting nature of an appreciation of beauty – Julia's fascination focused on Keats' belief in the power of poetry. But she played down the plaque's significance as more of a practical solution, screwing it into floorboards in the living room to cover a large hole. It was, she said, 'partly because it would fit and partly because I liked it', adding that she didn't 'really think about it' and didn't 'even tell people about it when they come in', despite its prominent place in the middle of the room: 'It's not meant to be a proclamation. It's meant to be – a *note*. A note to guests or note to self, whatever, you know. Something like that'.

The fact that there was only allowed to be *one* museum publically marking Keats' life in *one* of his houses created, as Julia pointed out, a simplified, undiluted public heritage experience. But to Julia's eyes, this signalled an absurdly random form of public display – granting some places more significance as 'heritage' than others as if one of Keats' homes had actually been more significant to *him* during his life; these 'other' homes needed to remain hidden *in order* to maintain the significance of the one earmarked as a site of heritage. In defiance, Julia had created her own plaque, giving this the dual purpose as also a *functional* element of the home – doing a job in covering a hole. As the home was *not* a museum, not deemed an ossified version of a particular past, she didn't have to 'think about it' too much; reinforcing the home's feeling of ordinariness – its assumed lack of significance as a heritage site – allowed her the freedom to get on with the daily round of everyday life undisturbed.[2]

Julia's intervention complicated assumptions about memorials as only ever *public* displays. Here, the private commemoration of Keats' life did not seek or require an audience – like the placenta tree and cat's grave elsewhere. But, in a similar way to the art books removed from Otis' flat in the east London square, the plaque also symbolised a quiet defiance against official 'values', replacing these with ones of heartfelt and respectful senses of relatedness.

In contrast, Gemma had not bothered to seek permission to create a memorial on the front wall of *her* cottage, believing this to be her *right*. Indeed, for her, to remember was an *ethical* act. Her plaque commemorated a famous horse that was buried nearby, owned by a 19th-century colonel who rode it into a number of battles. After the horse died, the colonel had buried it in the corner of a field a few feet from Gemma's cottage, erecting a memorial to it which was later taken to his private garden when the field was sold. By removing the stone from the site of the horse's burial, Gemma believed this 'disrupted' history and had wished to make amends:

> I thought it was a shame that the horse is presumably still there ... So I carved all the words off the original monument [copying a photograph in a local history book] ... because I think it's important that these things shouldn't be forgotten. And by moving the monument he – sort of – *disrupted* history.

Gemma carved the words – a description of the horse's life and a poem by the colonel – onto a huge piece of slate and nailed it to the front of the cottage, thereby reinstating the link between the place of its unmarked remains and its memory: 'I just thought, if he's there, people should know about him'.

Beyond the plaques, some participants showed other more private forms of display within their homes to commemorate past residents. On a shelf in the living room, Julia had a sketch of Keats in a photo frame next to another photograph of a group of friends on holiday. Adam had a framed poster advertising a music hall performance with his star Harry listed on the bill. The poster, he said, would stay in the house if he moved – another form of object to be passed on. Ben also displayed an enlarged, framed version of the servant's advert he had found in the book. Cathy had likewise enlarged, hard-backed copies of chosen examples of the famous photographer's works, placed above the fireplace in her bedroom – here an acceptably discreet, private celebration of his presence in the home. Lastly, Brian showed me two small decorative figures – a pig and a tortoise – sitting on a window ledge. He had purchased these from the showroom of Garrard the Crown Jeweller, having discovered that the original Garrard lived in the previous house on the site. 'I'm creating my own inheritance', he said: 'Putting back something of the Garrards ... Keeping the spirit – in this house'.

Thus whilst many participants focus their interest on the ordinary past residents listed in the census records (Cannell 2011), the more tangible evidence of past lives, afforded by default of a more public profile, allows for particular expressions of the home's heritage – the various objects, photos and paintings displayed on the home's ledges, shelves and walls. This further intermingling of house and personal histories extended people's own stories in and of their homes; the act of acknowledging these past residents as part of the more personal curation of senses of belonging through the display of material cultures is far less a negotiation here, far more a *celebration*.

Figure 7.1 This original door handle in a living room became a focus during one
　　　　guided tour.

Conclusion

These objects signal different ways in which people identify with their home's
past than reflected in the edicts embodied by the official arbitration of heritage 'value', reflecting broader ways in which participants personalise their
relationship with the past – sometimes in defiance of or as a counter to official
priorities and requirements. The latter decisions concerning what is allowed to
be more public (or public-facing) or relegated to the private sphere reflect an
additional set of contested values beyond the already-complicated responses
to the more public and private spaces of the domestic *interior* – in turn
reflecting attitudes towards and practices around what gets to be displayed or
hidden in which spaces and rooms.

　　A focus on the home's history also leaves many participants to reflect on
their own place in its 'pantheon' of inhabitants, both in the past and future,
and this in turn leads to a sense of responsibility as the home's temporary
caretakers or custodians. For some participants, the home itself, as a distinct

object or agent, needs to be shown respect; being a custodian embodies a desire to act correctly in making choices about the home's inherited material fabric. Whereas this is in part motivated by a desire to put one's own *personal* stamp on a home, for many this is an ethical concern, to improve a home for the future, passing it on in better shape. This, however, is less possible for those who do not own their home and therefore are powerless to protect or improve it. Here, those without the money or the ability to make changes – such as council or private tenants – are left to recycle, reuse or 'live with' the inherited past. In some cases major maintenance issues create a more mal-evolent presence of the home's past – the 'cash-eating monster' feels, to some, that it will eat them up.

Despite such limited custodial choices for some, for many participants the question of how one might improve a home rests on whether to restore it to an original state, or whether the home's identity is deemed more pro-gressively to be a story of change over time. Either way, in most cases, there are compromises to be made, between a desire for a home to be comfort-able and reflect personal tastes, and a need to respect or emulate its history as central to its 'identity' or 'integrity'. This makes it harder to define the value of being a 'custodian' of a domestic home one lives in, mediating as it does between respecting others and catering for one's own needs. Most people were aware and reflective of the choices they made in relation to broader cultural understandings of the value of the past. But the key issue remained: how to value 'the past' in relation to beliefs about how this informs a home's 'identity'. This dictated what people decided constituted the eth-ical act of 'improvement', further complicated by acknowledging the rela-tive values granted to the aesthetics – and (although less linear), changing aesthetic *tastes* – of different past eras. Whereas participants were often, as described, quick to judge the most recent resident's material changes, they are aware that others might also judge *them*. Thus in deciding what might or should be made to *continue*, being a custodian required mediating between personal beliefs about the nature of value, an appreciation of the home's original aesthetic, and an understanding of what of necessity *changes* and should be allowed to do so.

In part, people found a compromise in judging how to negotiate such home-making choices by applying the flexible category of what was deemed 'in keeping' or the right 'fit'. This reiterated common ideas about good design, in one version constituting

> how things fit the hand, how furniture fits the body, how people fit in buildings, and how buildings fit the landscape ... finding this sense of fit between people, places, and things. And if we think of design as being about fit, we consider not only the physical dimensions, but the moral and social ones as well.
>
> (Busch 1999: 25)[3]

How people accommodate themselves to pre-inhabited homes, and in which ways they are able to make *themselves* 'fit' materially, morally and socially, also requires an acknowledgement that they may not always be in control of such choices; that the home is likely to, in any case, have the last word.

Notes

1 For an examination of the mantelpiece as a form of domestic biography, see Finn 2009; Hurdley 2013.
2 The national heritage body English Heritage offers local and regional authorities guidelines to help decide which of any number of homes a historical figure lived in should be chosen for a commemorative plaque – the assumption being that there should generally only be the one. Criteria include not only how long the individual lived at an address, how 'productive' they were or whether they produced any 'notable works' whilst there, but also – rather more tellingly – whether they were 'happy during their time at the address' (English Heritage n.d.: 66). The latter raises problematic issues around the nature of 'happiness' – how it is defined or named during different historical eras (Dixon 2012), how it might differ across a lifetime, or, indeed, moment-to-moment. It returns us to ideas about how far we can access others' feelings, and what's at stake in assuming that past others at home were happy. But it also offers further insights into the role of emotion in 'official' heritage methodologies for judging the competing values of historic sites. Indeed, this rather reductive stipulation to consider a particular, positive emotional engagement with domestic places as a marker of significance suggests a bias towards memorials as *celebratory*, which may in turn preclude other more complex engagements between people and places and their affective economies (Ahmed 2004). Certainly, any argument for officially memorialising Keats' time in Julia's house, harnessed to such criteria, would fail this, as well as the other tests.
3 This includes gender, reflecting, for example, the ways the 'relationship between social practice and architectural form embodies varying degrees of gender segregation across time and space' (Llewellyn 2004: 44; Domosh 1998; Spain 1992). Participants' sympathetic understanding of the way that *kitchens* – previously considered the exclusive space for women and servants (often also women) – may or may *not* have created comfortable spaces to work in, is matched by a desire to have *more* control over such spaces, both in terms of function and aesthetics. Indeed, it is telling that Brian and Megan, showing the original plans they had been given by the architect's wife after moving into their 1950s house (built less than ten years previously), pointed to the 'tiny' kitchen. Megan was scathing: 'The kitchen, as always, was far too small. You could tell it was a *man* who'd done it, you know'.

References

Ahmed, S (2004): *The cultural politics of emotion*. Edinburgh: Edinburgh University Press.
Busch, A (1999): *Geography of Home: Writings on Where We Live*. New York: Princeton Architectural Press.
Cannell, F (2011): 'English Ancestors: The Moral Possibilities of Popular Genealogy'. *Journal of the Royal Anthropological Institute* 17.3: 462–480.

Dixon, T (2012): '"Emotion": The History of a Keyword in Crisis'. *Emotion Review* 4.4: 338–344.

Domosh, M (1998): 'Geography and Gender: Home, Again?' *Progress in Human Geography* 22.2: 276–282.

English Heritage (n.d.): *Celebrating People & Place: Guidance on Commemorative Plaques & Plaque Schemes.* Available at: www.english-heritage.org.uk/siteassets/home/visit/blue-plaques/propose-plaque/commemorative-plaques-guidance-pt1.pdf.

Finn, C (2009): 'Old Junk or Treasure? Every Item on This Mantelpiece Tells a Story'. *Guardian.* 3 January. Available at: www.theguardian.com/lifeandstyle/2009/jan/03/mantelpieces-family-history.

Goffman, E (1990 [1956]): *The Presentation of the Self in Everyday Life.* London: Penguin Books.

Hurdley, R (2013): *Home, Materiality, Memory and Belonging: Keeping Culture.* Basingstoke: Palgrave Macmillan.

Llewellyn, M (2004): 'Designed by Women and Designing Women: Gender, Planning and the Geographies of the Kitchen in Britain 1917–1946'. *Cultural Geographies* 10.1: 42–60.

Robertson, I (ed.) (2016 [2012]): *Heritage from Below.* New York: Routledge.

Spain, D (1992): *Gendered Spaces.* Chapel Hill: University of North Carolina Press.

Yarrow, T (2019): 'How Conservation Matters: Ethnographic Explorations of Historic Building Renovation'. *Journal of Material Culture* 24.1: 3–21.

Yarrow, T (2017): 'Where Knowledge Meets: Heritage Expertise at the Intersection of People, Perspective, and Place'. *Journal of the Royal Anthropological Institute* 23.S1: 95–109.

Conclusion
Heritage in the home in wider context

This book offers a fine-grained account of how one group of self-selected individuals, in the context of contemporary England, describes the ways in which the pasts of their homes are valued, what is experienced, explored or avoided of those pasts, what is believed, known, felt or imagined, and how these encounters impact on processes of making home. This is an overlooked aspect of heritage research, which has tended to focus on public sites rather than how people 'use the past in their everyday life' and 'make sense of and engage with the past ... across different contexts' (Moussouri and Vomvyla 2015: 99). By exploring affective, emotional and embodied encounters with domestic heritage, I have offered insights into how the home's pasts inform, trouble or enhance the ongoing project of securing subjectivity and senses of belonging.

The home as a site of heritage practice, experience and knowledge becomes a historical object in its own right, a distinctive presence full of many space-times, one which might suggest 'lasting, secure depth, fixity: the character attributed to particular tourism or heritage sites' (Crouch 2015: 187) – or at least reflect the *desire* for uncomplicated senses of rootedness, continuity and stability. Home as a space of heritage also becomes a domestic archive full of the traces of others, not a passive, inert backdrop to events but co-constituting the various relationships which emerge within it. Exploring what people do with the materials and objects they inherit suggests how they 'curate' their homes in ways which negotiate between their own memories and aspirations and those of other past and future residents. Tim Cresswell's (2012: 175) call to explore 'other kinds of collecting and other kinds of space as archived, including places themselves' here finds an apposite example, one which goes beyond the normal association of collecting practices within the home as a reflection of purely *personal* memories and histories (Woodham *et al.* 2017).

Self-other relationships at home

The book has built on my previous study of experiences of the domestic uncanny to offer insights into home as a place which, rather than as traditionally conceived offers privacy for selves seeking sanctuary from external 'others', is rather a space which is shared by a range of selves and others over the span of its times, encountered through the home's material spaces, through imagined or more-than rational experiences, through archival research, and through engagements with living past and future residents. This sharing of private space suggests ways in which a range of others have, to more or less degrees, power to affect senses of belonging and practices of making home, offering a distinctive context to examine Sara Ahmed's (2004) framing of relationships between selves and others in terms of affective economies – the dynamics of power dictating how selves are created by others.[1] Specifically, it engages with the temporalities – in particular the pasts – in which such relationships emerge and are enacted and negotiated in the present. In turn, the exploration of temporalities is harnessed to the way these relationships are gathered in within the specific geographical context of home.

The continuing desire to create home or be 'at home' – to align the spaces and social relationships within home with a sense of a self 'at home' – sets up particular tensions in dealing with others who have been, imaginatively or through their objects, inherited rather than chosen. The degree to which the presence of others at home is tolerated, welcomed or rejected depends on people's attitudes, beliefs and values as much as their power to act. A focus on the past at home offers a vantage point for exploring a range of encounters which, at one end of the spectrum become ways others are strategically made similar to self, whilst, at the other, are contained, distanced or stripped of their 'otherness'. For some, however, the process of making home is one in which they themselves come to feel like 'others', in the context of, for example, intransigent owners or, as will be explored, the wider socio-economic context. Just as people choose the 'pasts' they can relate most to, they also (where they are able) choose homes and future residents. Attempts to feel 'at home' can exclude others who are *too* different or threatening, whilst enjoying the fascination and frisson of difference at a safe distance. In turn, memories which are not one's own can be colonised or evicted. There are also many examples of a desire for connection not as a seeking after the self and same but as a means of understanding or empathising with other lives, be they servants from the earlier past or more recent residents come to visit (Landsberg 2004). Either way, people had a high level of awareness of their own sometimes complicated responses to encounters with others, such as, for example, a desire to communicate – including leaving clues to one's own identity for the future – and a simultaneous desire for privacy. Exploring both imaginative relations with earlier and future residents and more immediate ones with recent and current ones suggests how the attempt to gain senses of belonging to home is an unstable and ongoing project. The ways in which the past is felt or made to be closer or more distant, how the networks of past others are pacified, negotiated with or wilfully ignored, the perception of similarities and differences, and what forms of accommodation or exclusion result, suggest that the process of securing subjectivity can be enhanced or challenged by the 'presence' of past others – but never complete.

Custodianship and pride in work is, in turn, a complex response to home as a responsibility to others, a passing on of one's own positive contribution, and a desire to create a home to one's own personal taste and comfort. Tensions, compromises, and feelings of guilt about choices made suggest, again, this process often continues throughout people's lives within their homes. The past at home is also engaged through embodied practices, through these acts of home-making such as renovations but also in terms of the everyday habits and rituals of making home and being at home. The Bergsonian forms of embodied memory and 'habit memories' (Young 2005) suggested in such accounts draw attention to the ways experiences of the past emerge through haptic and visceral engagements with the home's material inheritance, expressing a particular form of affective intensity. Like the frisson of touching original materials, this also became a way to reinforce benign continuities and connections with past residents, reiterating assumptions about the feelings of familiarity gained through repeated everyday routines – here incorporating the routines of past *others*, their own repetitions over time in the same spaces creating a collective sense of embodied presence and belonging.

This focus on embodied memory might foreclose considerations of broader contexts, stripping such relationships with others of their social complexity. In some ways, this is the point of them – to compartmentalise more complicated relationships with others. But embodied encounters with the 'memories' of others can also become a bridge between the personal and social rather than ignoring the social context, expressing the more intimate moments in which broader relationships emerge. Those who had less power to act tended to have the more challenging experiences of the past, whilst for others, imagining or knowing the physical deprivations of past lives cannot fail to force an acknow-ledgement of disparities and reflect on the assumed privilege of their own lives. Whilst the past can be cherry-picked and interpreted in ways which reinforce particular normative and stereotypes – a strategy which keeps past residents at a distance whilst enhancing senses of familiarity – there is also the possi-bility of being forced to deal with past others who are not the same as oneself, particularly in relation to encounters with the more recent residents or those for whom more details are available. Sometimes people took a defensive pos-ition in responding to particular differences, reflecting their own prejudices or expressing anxieties about *naming* and therefore acknowledging them. Yet, again, engaging with past others could also be enhancing, and offer a means by which home could be made a place of hospitality and care.

The assumption in the literature that in order to make home a 'familiar' space one needs to banish the past is at times sustained through strategies of distancing – at an extreme, including forms of 'exorcising' (Bachelard 1994; de Certeau 1984; Miller 2001). But this is not the case in all accounts; the various presences of the past are subject to different forms of negotiation. For many the home's pasts are intrinsic to the identity of the home and, in any case, as shown, elicit a wide range of responses. To describe the impact of encounters with heritage at home as either enhancing or alienating would not be to reflect the complexity of these engagements. The ongoing dance of negotiated self-hood suggests that the wider ethical project to seek purposeful conditions for the development of empathy, contact and connection with others needs to be explored in more detail. There is a significant opportunity to understand these relationships with others, adding to emerging work on the affective politics of more conventional sites of heritage.

Affective economies in the home

As shown, people also have differing degrees of power to act, although this doesn't preclude a sense of belonging, respect or protectiveness towards their homes, sometimes deemed a physical presence in distress. Rituals of 'making one's own' or 'moving on' are also less available to those forced to live with or recycle objects – here the sense of sharing with strangers is palpable, imposed by circumstances.

Official ideas about the value of the past also imposed constraints, limiting some people's ability to make their own choices in negotiating or respecting the home's past whilst evolving their homes for the future. In this example, everyday heritage practices and beliefs can be seen as a direct counter to official discourses. But in other ways, as shown, we should avoid assuming that everyday places and encounters are by default the site of counter-hegemonic practices; they bring a range of values, beliefs, needs and prejudices to the home-making project which cannot so easily be framed. Many of those who own their properties also, in different ways, grant power to others – for example, as described by social contexts in which edicts dictating which objects must be left in situ are accepted (an interesting parallel to edicts dictating which *buildings* must not be modified; both attempts to curtail change). The granting of agency to the home itself also reflects how far people feel the need to respect homes as containers of accumulated other lives in need of accommodation.

Elsewhere, the value granted to found objects suggests a respect for the original owners which again, in part, dictates how these objects are handled. In terms of more recent residents, many have power to pass on an 'inheritance' – gift-like or curse-like – which enhances or disturbs senses of being 'at home' – challenging ideas about what and with whom we intimately relate by extending the traditional contexts in which legacies are passed from one generation of family to another (Owens *et al.* 2010). Indeed, if forms of 'inheritance' described here create the backdrop to forms of relatedness between different selves and others, beyond a traditional narrowly-familial focus – and indeed we define 'inheritance' in relation to the range of subjects and objects involved in acts of inheriting – this raises several questions at the heart of this book. What do we *choose* to inherit? What ethical positions can we locate in such choices to act? Or, put differently, what degrees of choice or constraint do we have in how we *manage* what we inherit? If inheritance is about what lingers of the *past*, it is also about *passing things on* to the next generation, requiring acknowledgement of the complex temporalities involved in these different forms of exchange between selves and others. Lastly, in defining inheritance as a backdrop to the forming of relations between different selves and others, how do we account in our theory for the fact that inheritance is often regulated through *emotive* and *imaginative* responses – the things, beliefs, forms of knowledge and intangible experiences emerging from what is 'inherited' in the pre-inhabited home and passed between residents?

Spectral heritage in the home

More-than rational beliefs about the past's presences, atmospheres and agencies were, indeed, extremely common – surprisingly so in a project in which this had not intended to be such a prominent focus of enquiry. This suggests, of course, that we need to take such forms of engagement with the past more seriously, given their strong impact on people's emotional responses to their home and figuring of its pasts. In turn, the relationships between knowing, imagining, experiencing and believing were interconnected. Is it possible to engage with the past without recourse to acknowledging the different configurations of these responses? Embodied encounters with the material 'otherness' of past residents also reinforced ways in which the material residues of the home become uncanny, being both intimate (even at times banal) and yet incompletely available or knowable – reinforcing ideas about the uncanny, the strange within the familiar. These encounters support an argument for the extension of experiences and practices of 'intangible' heritage to more-than-rational beliefs and experiences, including as here their contemporary Western context – what might be called a form of 'spectral heritage'. We also need to study those domestic 'rituals' many people were engaged in, such as objects left in situ or those knowingly left behind or hidden for future others to find. These, arguably, are just as much forms of evolving 'intangible' cultural heritage as those officially listed and earmarked for safeguarding, and indeed they speak back to the incompleteness of the latter project. What are we to do with such quasi-private, ad hoc and often spontaneous 'traditions' which seek to both protect people *from* and connect people *to* each other over time? These are no less valuable forms of 'living' heritage, and, arguably, they are in some ways more valuable, not just because they are often overlooked, but because they appear to be so ubiquitous. And, unlike the tendency to focus on forms of intangible cultural heritage which reflect narrowly-defined cultural and familial groups, as described, the 'traditions', beliefs and practices described in this book offer the greater challenge of accessing how people 'communicate' with a pantheon of others, across time, in inheriting things, memories, atmospheres, stories – and passing them on.

This book, thus, contributes to a growing acknowledgement of the need to attend to the 'subjective qualities … such as atmosphere, spirituality, feeling' (Jones 2010) of encounters with heritage sites. Watson (2018) also argues that 'official and private meaning making' might be 'very different from each other', referring here to 'folk tales' and other 'popular narratives' when discussing engagements with heritage. But we need to take care that acknowledging these forms of 'meaning making' do not reinforce the binary between experts and the rest of the 'people' – the assumption being that such popular practices are the preserve of the less informed or educated (Bennett 1987, 1999; Bennett and Bennett 2000). What (surely) gets missed – what experiences, encounters and histories remain hidden – when these forms of engagement with the past are ignored or placed too quickly into particular categories of experience?

Home as a space of heritage, thus, becomes a place which is not mundane or ordinary and cannot be so easily bracketed from those things and places which are considered *not* ordinary or everyday, whatever and wherever they might be. The way the so-called 'ordinary' pasts of homes are infused with atmosphere and agency suggests that it is not just the charismatic places and objects with 'time depth' which are granted extraordinary presence.

Negotiating continuity and change

A focus on domestic heritage also draws attention to the changing rhythms and textures of living within homes, of getting stuck, of moving on. This reflects a further key theme, emerging from the qualitative material: how encounters with the past are expressed through perceptions of *continuity* and *change*. Ideas about what stays the same or changes between *now* and *then* become part of perceptions of similarities and differences between selves and others, a means to keep some near and others more distant. Within the material spaces of the home, this relationship emerged as a negotiation between making changes and retaining older elements. For some, changes were made early on, allowing people to bed into their home, or to learn to work around elements which didn't fit so well, whilst for others modifications and repairs were made slowly, in a more piecemeal manner as and when money or time became available. For others, as described, the power to act was more limited, the imagining or knowing of the home's pasts narrowly shaped by the urgent needs of the present, the home becoming a burdensome, entropic, presence.

Questions about what can, should or do *change* and what can, should or are preferable to stay the *same* underpin home-making practices and the negotiation of senses of belonging. But these changes and continuities within the domestic interior are also impacted by wider changes and continuities *beyond* the interior. To conclude, I will reflect on how we might extend an exploration of the past at home to consider people's encounters with pasts, presents and futures beyond these material spaces of the domestic interior. Indeed, home is experienced at different scales – it is both an interior with its more public, private, surface and hidden spaces, but also it is part of extended senses of home as neighbourhood, in turn as part of broader everyday and transnational spaces (Blunt and Sheringham 2019). A focus on embodied memories within the micro-spaces of the domestic interior does not preclude an engagement with broader social memories and socio-economic contexts, as these have been shown to have an impact on people's sense of belonging to and at home. Below I reflect on the specific ways in which participants describe the impacts of wider contexts in which the relationships between change and continuity play out – geographically, socially, economically and politically.

Heritage at home in the neighbourhood

Various spaces *outside* the domestic interior seep into participants' narratives of home, as described throughout this book. For some, the home is quite literally porous, filled with lingering sediments from the past that have made their way in from the outside. Others look out of windows, noting change and continuity in the street, such as the old trees, or describe walking their dog past a previous home, the memories of which influence responses to their current one. Some believe their home's history needs to be understood in relation to its local contexts, or are given local information by long-standing neighbours, often emerging from brief conversations or meetings rather than through formal research. As Joy said about her chats with a neighbour who told her about the past residents of her flat: 'It is just having a good gossip really about people what lived here'. For a few, local history is a focus because their home is not deemed old enough to be interesting, whilst elsewhere it is because their home is considered intrinsic to its environment – such as Pam's description of her converted farmhouse's unfoldedness within its remote valley. For others, discovering the close proximity of colourful or significant past events reinforces a fascination for what has changed locally – soldiers hanging out of the windows of a local house to declare the end of World War I or, in Dan's case, the discovery that his east London garden had been part of the public 'pleasure gardens' of an old inn: 'This used to be the taking-off field for hot air balloons, which was a Victorian play-time activity. That's quite an interesting thing. My garden used to be a hot air balloon take-off place!'

The neighbourhood itself becomes an extension of home through repeated embodied routines and practices, allowing connections with past residents, such as Yolanda's ability to see the locality through the eyes of the first resident as she takes the same shortcut to the shops – the knowings and imaginings of the local past interconnecting with people's own senses of belonging through feelings of familiarity. For Martin and Carol, this has evolved in line with their local knowledge:

Not just in your home, but you feel at home within an environment. You know automatically where the dry cleaners is and the chemist. *Familiarity*. I mean I used to always, *always* – for the first pretty much year – take the wrong turning driving home. I'd get to the mini roundabout and turn right and I needed the next turning. So these things that you do when you live somewhere for a period of time that become automatic. That threw me as well for a while with moving, you know.

(Carol)

Neighbourhoods are, as Kathy Burrell (2016: 1603) suggests, difficult to define, and 'may represent different spaces' for people. For many participants, home as neighbourhood interweaves with home as interior in different ways. Through these various spaces, people track continuities, explore what has changed, or make places familiar over time. In the earlier example, a sense of belonging emerged at the point when *not* making a wrong turning became 'automatic'.

Nostalgia

Popular interest in history is often assumed to be in part a nostalgic escape or, at the least, a form of complaint about the present. Here, sensitivity to the relationships between what continues and what changes is deemed to reflect anxieties about change itself. Australian historians Paula Hamilton and Paul Ashton (2003: 2), commenting on the findings of a public history survey, observed an 'increasing obsession with the past both personally and in a range of public forums', a 'growing preoccupation with the past for public consumption', and the 'booming public interest ... in local history and family history research'.[2] They argued that this is a reaction to broader 'contemporary challenges to cultural identities, social authority and institutional shifts within a context of globalisation and rapid technological change'. This dovetails with commonly-expressed views about nostalgia (originally a literal 'home sickness') as a 'general longing for past experiences that allow for the maintenance of identity in the face of changing circumstances ... a sentimental longing for the past' (Hibbing *et al.* 2017: 231). At worst, nostalgia can lead to a 'falsification of history' (Strangleman 2013: 28; Boym 2001), such as in the way that ruins can be aestheticised, stripping them of their socioeconomic contexts. But others have attempted to reappraise nostalgia as a more productive state (Blunt 2003). Vanessa May (2017: 402) describes the act of recalling past places associated with senses of belonging – or 'belonging from afar' – as a way of 'mobiliz[ing] the past in ways which can enliven the present', with nostalgia allowing for a 'sense of continuity of self even under changing external circumstances' rather than being an 'uncreative form of conservatism' – a 'harking back to an idealized (and partly fictionalized) past'. Elsewhere, Tim Strangleman (2013: 28) suggests the 'radical or oppositional aspects of nostalgia – where knowledge of the past makes a dialectic intervention in debates about the present', and Alastair Bonnett (2016: 16) adds that nostalgia is 'rarely articulated in terms of a past that was simply or completely better or gone', drawing attention to the 'politics of loss and, more broadly, attachments to, and uses of, the past and present' (17). This, he adds, includes the persistence of 'loss and longing' amongst migrant populations (97). Bonnett argues for a need to 'move on from the *a priori* categorisation of nostalgia as irremediably passive, conservative or uncreative', and to 'dispense with the discourse of instrumental judgement, of scolding people who turn to the past for lessons, models and inspiration'. Instead, we should 'draw attention to the ambivalent, necessary and contradictory sense in which nostalgia resides at the heart of belonging and attachment' (16).

Some participants do appear to imagine the past as somehow better, or more interesting, than the present, but it is rarely so simple. Here nostalgia is used as a signifier of a specific complaint or preference – the love of a past era's house design or aesthetic features enhanced by the dismissal of modern (usually post-1960s) ones. Respect for the assumed superior craftsmanship and individuality of hand-made rather than mass-produced elements of homes suggests nostalgia for the pre-industrial past, but also reinforces pride in a home's specific qualities and a desire to emulate an assumed ethic of patience and attention to detail. Thus nostalgia, as other forms of remembering, is part of a repertoire of ongoing intimate and affective encounters which trouble and mould subjectivities, reinforce values, create responsibilities, and influence home-making practices. It also allows space for the past to be *different*, offering room to ameliorate more challenging senses in which the past is felt to be *too* present. But vicarious feelings of nostalgia for past lives are also counterbalanced by a self-conscious awareness that one cannot completely control imaginative responses through processes of cherry-picking and compartmentalisation. And, as Pam pointed out, we might try to 'colonise' the past of the home, incorporate it into our own narratives, yet it also affects us back in ways we can't always anticipate or control – the home's past 'colonises us back'.

Nonetheless, some participants do seek solace in the past as a reaction to feelings of social or economic lack or anxiety in the present, and it is important to understand what lies at the heart of such feelings. How might, for example, these offer insights into what inhibits some people's search for a sense of belonging to home – such as feeling *oneself* to be an 'other' or 'othered' as much as encountering imagined past 'others' (or, indeed, creating forms of 'otherness' in order to reinforce one's fragile sense of belonging)? How far do broader social, economic or political *changes* impact on people's sense of belonging at home?

Local belonging and identity

In order to consider these questions further we need to locate them within theories of local belonging. Here we find nostalgia used as a derogatory term to signal a suspect political conservatism which is, in turn, associated with essentialised and parochial forms of identity – the role of nostalgia acting here as the negation of a global, fluid and future-oriented world in opposition to one which is fixed, inert, sedentary, and of the past (Massey 1995; Ley 2003).[3] In her influential essay, Doreen Massey (1991: 26) pointed to the 'search after the 'real' meanings of places, the unearthing of heritages' as being, in part, a 'response to a desire for fixity and for security of identity in the middle of all the movement and change'. Heritage becomes a way to reinforce places as having single essential identities, with the 'identity of place – the sense of place ... constructed out of an introverted, inward-looking history based on delving into the past for internalised origins'. In contrast, Massey argues, 'what gives a place its specificity is not some long internalised history but the fact that it is constructed out of a particular constellation of social relations, meeting and weaving together at a particular locus' (28). The recent shift of focus to mobilities builds on this idea. Caitlin Buckle (2017: 4), for example, describes how the reframing of place as linked to mobility has 'led to a more dynamic and inclusive perception of home as a lived experience that can be moved from place to place' – a useful contribution for reflecting how most people have ended up somewhere else than where they came from, whether far or near, forced or chosen.

But the mobilities turn is not without its critics. Jana Costas (2013: 1472), for example, points out that the 'celebrat[ion] of mobility as deviance, resistance and the displacement of fixity' has been criticised for 'over-enthusiastically embracing mobility as a desirable and universal condition' – its 'liberatory' potential rejecting essentialist/sedentarist notions of place, identity and meaning. Costas argues that this 'doesn't distance itself enough from a neoliberal stance on mobility, which similarly rejects any notion of place/stability ... and celebrates fluidity and movement – seen as emblematic of opportunity, choice, innovation and creativity'. Tim Cresswell (2010: 28–29) also calls out some mobility theorists for positioning themselves as progressive, incessantly focused on the present and future: 'It is *now* that is mobile while the past was more fixed ... the present and the future [are] about the mobile and the dynamic while the past [is] about stasis and stagnation'. Grafted onto such an idea is a geography of nostalgia which assumes that a focus on local identity and belonging is something retrograde – 'local' here standing for rather different values than those counter-hegemonic claims rehearsed in the introductory chapter.

Geographer Hayden Lorimer (2013: 183) – whose work carefully explores changes and continuities in the landscape – points to a further shift to what he describes as a 'new sensibility of critical sentimentality', a 'vein of romantic antiromanticism' where

> the promise held in the idea of a return is something very different from shamelessly going back to the future. Rather, it is a stylistic turn, raising possibilities for writing subjectively and affectively about what is to be cherished in vanishing landmarks.

He concedes with self-irony to being as 'susceptible' as the next cultural geographer to 'landscape nostalgia' –

> Arguably a little more so; finding comfort in the old rather than the new, and prizing landscape's historical associations. I can tell when the pangs are coming on, to a point where parody often seems like the best antidote, or the easiest way out.

This honest reflection on a trend in academic landscape writing was written in the context of repeated warnings: that what stands for 'locality, custom, tradition' should not be defended against the encroachments of the modern; that the 'existential and romantic power of ideas of place, *genius loci*, dwelling' might be 'nearly irresistible' – but we need to avoid them at all costs (Wylie 2012: 378).

Despite this, literature exploring ideas of belonging continues to place emphasis on the co-dependency of place, self and identity. Material theories of belonging focus on personal objects – useful for explaining how participants seek anchorage in permanent or original objects within the home, or in the slowness of some materials to wear down. (Indeed, the domestic display or archiving of *personal* objects might in turn become more important for reinforcing people's own presence within the home when having to also nego-tiate past *otherness*, such as dealing with found objects belonging to previous residents). Marco Antonsich (2010: 646) summarises how the 'emotional con-notation associated with belonging as feeling at home in a place' makes it

> not surprising that, at times, this notion is also rendered in terms of a sense of rootedness ... or discussed in relation to (if not even confused with) notions of place attachment ... sense of place ... and place identity.

He argues that it is 'generally agreed that feelings of belonging to a place and processes of self-formation are mutually implicated'.

Kathy Burrell (2016: 1602) also describes how 'belonging ... is at the heart of place attachment, something which is difficult to unpick but rests on a combination of physical, habitual and affective connections made between person and place'. Burrell's research into the effects of high population turn-over (or 'churn') in an inner city neighbourhood leads her to suggest that a focus only on place as a 'process' can be 'disconcerting for anybody who holds an emotional attachment to an imagined or remembered, and relatively fixed, *version* of their neighbourhood and what it constitutes socially and materially' (1604). She argues that although places 'may be theorised as ever evolving processes and interactions ... there can be an emotional disconnect between experiences of this fluidity and a human desire for settlement and certainty' (1612). She adds that

> transience and mobility are presented as being at odds with the mean-ingful construction of neighbourhood and belonging, socially and materially. A powerful dynamic of exclusion, otherness and blame is put forward, all oscillating around the impact of newcomers and movers, but the social and emotional dynamics of churn are more complex than this.
> (1607)

Likewise, Julia Bennett describes the assumption that continuity is 'central to a stable identity over time', whilst change is an 'unsettling, but inevitable, part of everyday life' (Bennett 2018: 449), her focus to examine how 'people negotiate the tensions between change and continuity in their lives' (450).

Burrell and Bennett usefully challenge us to think about the ways senses of belonging are continuously negotiated in our relations with others, and how these are perceived to change or stay the same. The sensation of 'stickiness' (as famously described by Jean-Paul Sartre as ambiguously situated between liquidity and solidity) offers a useful metaphor for exploring the state of 'mobility' as neither a 'threat to stability or recipe for emancipation from places' (Costas 2013: 1473). But if we consider places as neither fixed nor fluid, rather shaped through the tension between the two, we can follow Ahmed (2004) in asking what gets stuck to whom and how. Added to this, we might ask what hopes are invested in objects and places (such as homes) being allowed to be more – or less – 'sticky'. How are some things or people made to stick more or differently to other things? Do people *get stuck*? What gets stuck *to* them?

These questions are pertinent given the majority of participants described in this book have lived in their homes for a period of time, in some cases many years or decades. It is useful to focus on the contexts of how and why they got to *settle* – perhaps an overlooked area of research – and, in particular, how this affects their perception of the passing of time in relation to their senses of place and belonging. Indeed, those who have lived a long time in the same home are never, arguably, in a state of uncomplicated stability but are continually negotiating their relationship to change. Within the home, for example, there are the impacts of changes over the life course and inter-generationally. And the longer participants stay, the more their own memories become imbricated with those of their homes and locales. Beyond the interior, whilst some people suggested they enjoyed a taken-for-granted feeling of belonging afforded by being settled and familiar, others complained of being stuck due to their socio-economic circumstances. People also became both a witness to neighbourhood change over time and felt impacted by it: an area's changing status – processes of gentrification or physical changes to local infrastructure; social change in relation to the area's make-up, support networks or degrees of neighbourly sociality.

Firstly, as described, the wider contexts are often considered intrinsic to their understanding of their home. Leah explained that her history research included 'the local area, the people, and then the style of the house', in order to get 'the big picture of it' – how the home was shaped and fitted into the history of the locale. Leah pointed out that ex-husband Alex – a past chair of the residents' association known for leading a successful fight against a local major bypass – had been 'the *daddy* of the street. You're like the *king* of the street'. Alex had moved to the house 36 years before as a young, working-class tenant. He eventually got the chance to buy the house for a low price:

> Remember in those days you couldn't get a tenant out. You were a
> protected tenant. And there was a German lady and her husband on the

top floor. They had been there for 30-something years, so he couldn't get them out. The basement was wringing wet [laughs]. So there was really only that floor … it was in a bad state.

He noted a shift in the street's fortunes from the days of the 'noise and the dirt' from the steam trains from the adjacent (now disused) track. This explained why, despite the size of the homes, the occupants never stayed long. Since the closure of the railway line, the street had become a stable, middle-class enclave: 'They come and say they are going to stay here for a bit, 'Oh, we are in transit, we will stay here a couple of years'. Yeah? – They *stay*. It's such a great place to live'.

Others described how perceived local changes reinforced anxieties about their own future in their homes. They were already aware of their own relative transience – even those who had lived many years in the same place – in the light of their home's wider temporal scale. Knowing the history of their home did not necessarily lead to feeling more rooted but reminded participants of the temporary nature of their own circumstances. And whereas the value given to the home's past is often viewed in relation to the degree of time between *then* and *now*, this is matched by the value given to the length of time *past* residents spent in a home. For most, this included the common assumption that senses of belonging could be mapped onto length of residency. For example, as described, census records from the mid-19th century showed working families staying for relatively short periods of time, perceived by many participants as a marker of poverty or hardship – in contrast to the assumption that the ability to be mobile is a privilege of the Western urban middle classes (Morley 2000).

For Sara, the census records, reminding her of the transience of past lives in her home, made her reflect on her own response to change more than she had 'allowed herself to think'. Like others, Sara had an interest in contextualising the history of her home as part of the local neighbourhood, and she pointed to her long-standing neighbours as a source of knowledge as well as friendship and mutual support. One friend who had lived in the road for decades had, for example, told her 'about the interesting or weird people who lived here back then'. But she was anxious about impending local changes, including plans for a new 'super sewer' which would make her street – already close to an industrial area – the main cut-through for heavy commercial vehicles, increasing noise and pollution. She said: 'The disruption in terms of what they are proposing to build and change is just enormous', adding that knowing the history of the area had offered 'pause for reflection' on the fact that 'things are ever-changing. Whatever it's like, whatever it's been like, whatever it's like now, it's not going to stay like that'. This wider perspective offered a means of coming to terms with what was happening around her, looking back as a way of mitigating anxiety about the future. But other changes also concerned her. She had become 'entrenched' in the area, having made friends with some neighbours and enjoying everyday encounters with local acquaintances: 'You see people you know even if you don't *know* them. You know them by sight. I think that builds into the sort of fabric' – a mixed blessing, as not all her neighbours had been 'nice'. But with increasing transience in the area and some of her friends moving away, this 'fabric' was breaking down. And as a recently-retired woman living on her own with arthritis, she worried that she would eventually need to leave her flat. She said: 'The other *horrific* thing is that, you know, I couldn't be here any longer because I couldn't cope with being here. I certainly don't want to be thinking about that. Because that's *too* horrible'.

Tenancy and transience

This mixture of changing personal and local circumstances impacting senses of belonging to home reflected the wider impacts on some participants of, for example, government housing policies. A few participants benefitted from policies which had allowed them greater choice in improving their circumstances. As described, some of the older participants had previously benefitted from having 'sitting' or protected tenancies which could lead, as in Alex's case, to buying properties cheaply. Lillian, having moved into an unfurnished flat in Brixton as a sitting tenant, also eventually bought it for a cheap price. Elsewhere, people described moving between public and private housing. Yolanda was bought up in 'poor accommodation, a Victorian clearance area' with no hot water and an outside toilet. The family was later moved to a 'very nice council flat but in a not very nice area'. Susan likewise was brought up on a council estate but cut herself off from her family, as described, eventually buying her own house in a different area of the city. William's downward mobility was a direct effect of his prison sentence many decades previously, leading him from a comfortable life in his own house to a council flat. For him, imagining himself in the previous manor house on the site had become a compensating fantasy, taking himself away from his 'ugly' studio room: 'It made me feel grand. ... This research *lifted* me'. His feeling of isolation in the flat – compounded by the recent death of a friendly neighbour – led him to plan his future escape: 'No, I can't end up like this'.

More recently, the UK government's *Right to Buy* scheme has had a major impact, benefiting some participants but not others.[4] Alice, for example, was an early beneficiary, cashing in on her council tenancy in a flat within historic east London almshouses, and making enough profit to buy a large house 'in a very, very rough state' off a man who had also bought his house with the *Right to Buy* scheme. Rita also bought an ex-council flat off the previous tenants, who had likewise bought it from the council. Julia, however, was against the scheme, believing that someone else should benefit from her council tenancy now her economic circumstances had improved.

For those participants who struggled – council or private tenants, those who had bought their council or co-op flats or couldn't afford to do so – the burden of maintaining old properties was often combined with a growing sense of isolation as the social mix of the neighbourhood changed with council tenants cashing in and moving out. Lydia, as described, had felt pressurised into buying the tenancy of her co-op flat – the discount was being reduced – and became burdened by severe maintenance issues. Here the history of her home was, as she said, 'the history of the house *cracking* – the history of the structure of the house. You're lying in bed and there's a new crack in the ceiling, and that crack kind of – *grows*'. Here, the *Right to Buy* policy affected the social mix of the area as people sold on and moved out, impacting the original communal and supportive principles of co-operative living: 'Lots of people have moved on, lots of people have moved out ... lots of funding has gone ... the community centre is gone. ... Funding was taken away from the youth provision'.

Elsewhere, Elaine continued to live in the council flat she had shared with her parents, both now deceased. Being too poor to buy the flat, she felt increasingly isolated and her flat was neglected by the council who she felt was trying to force her out. She described the irony of the block's original middle-class roots – it had been built for Edwardians wealthy enough to have bank accounts, and later sold to the local council. She had found out that her own flat had been lived in by a gentleman with a cook and valet, and felt more interest in the servants, wondering where they would have slept – noting how she herself would have been a servant if alive back then. Now the flat had become 'squalid'. She pointed to black mould in the corridor, the ceilings falling down, damp peeling walls:

> I've had four or five surveyors come in and look at it. And all the builders that have come in have said, 'No, I'm not touching it – I don't want to do it' ... The place is rotting, and I don't have the money to do what I'd like to do with it. I'm relying on the council and they don't care and they're not interested.

The council was trying to persuade her to move into a 'modern one-bedroom box'. She was holding out.

Like Lydia's co-operative, *Right to Buy* had, Elaine said, changed every-thing, leaving her feeling deeply isolated: 'People started going mental, they started like buying their property ... renting them on and becoming landlords themselves ... People got very greedy very quickly. We never did that. So the demographic has changed. It has now re-gentrified'. The immigrants who used to live in the block had moved on and mostly young professionals 'working in social media' had bought the flats: 'They sense that I'm different to them. They sense that I'm working-class'.

This feeling of social isolation is shared elsewhere. Miriam's boyfriend had bought his mother's council flat; she inherited it after his death. She said: 'I didn't really quite agree with it – we were obviously council tenants. It was the time everyone was buying their council flats. And so now I'm the *leaseholder* here [laughs]'. For her, interest in the past was a response to her unhappiness in an area where everyone was 'in a hurry, not knowing the neighbours. No one you can go and drop in on. That would be nice – going and knocking on somebody's door and having a cup of tea'. Her block had a high turnover of tenants, so 'you tend not to know your neighbours. There's a man living opposite ... I see him perhaps once a year. Very strange'. Her 'feeling about the past' and desire to imagine 'living in some past era' included a romanticised assumption that people living in closer proximity to each other would have enjoyed a greater sense of sociality than the vertical living spaces which had replaced the Victorian slums at the site.

Elsewhere, Sandra, as described, was also impacted by social change. Calling herself the 'last serving inmate', she was the longest-standing tenant forced for over ten years to sign precarious annual contracts for her flat in the charismatic but crumbling mansion – its fortunes severely curtailed by the liminal zone it found itself in, a motorway at the bottom of the garden. Sandra was an observer of all the 'comings and goings' of the 'transient population' – people leading chaotic or insecure lives who never 'live here long', including asylum seekers and immigrants.

Sandra was one of many examples of the inability of tenants to improve their homes because they had neither resource nor power. It was not that she did not *feel* a sense of custodianship of the house – she felt a strong sense of protectiveness along with respect for its history. Her dilemma was not, like others, about what choices to make, but about how to live with her *lack* of choice to act. The landlady's neglect of the property had led her to ignore and exacerbate its maintenance issues. There had been, for example, a major flood which caused 'utter chaos for a long time', and a line of mature trees at the bottom of the garden – which had acted as a shield against the motorway noise and fumes – had been chopped down for no reason:

> I do have a strong sense that this is my home, to want to nourish it and defend it and treasure it [but] it's constantly being attacked. ... The lack of appreciation of the culture of the building, the history, the original features. ... I would like the original things to be preserved and *valued*. ... [But] I am a toothless tiger.

These participants' feelings of isolation and frustration were also expressed as the physical discomfort of living in unmaintained homes, and a sense of being stuck – the inability to leave matched by a fear of not being able to stay. In such cases, there is a sense of failure as custodians, not because of a lack of care – they cared far more than the property owners. Their own sense of precariousness, of not being able to make plans for or imagine the future, does not however preclude their sense of belonging to home. Joy described this in relation to being a private tenant:

> I know what the building is now, I know what it used to be, I want to know what it [the land] was before. ... It gives a bit of continuity – while

I'm here it's *my* history, of where I'm – of what I'm living in, or what I'm living *on*. ... This house is part of me while I'm here. ... I want to know about the – ancestry of the house, so that is also part of me. ... I think this is interesting because it's attached to *me*. It's mine, where I live. Although it *isn't* mine [laughs].

For those who did own their homes, the feeling of transience in their neigh-bourhood also affected their sense of belonging. Sara described the changing social mix in relation to another trend – buying homes to rent out, or 'buy-to-let', a way of investing money. She felt that tenants didn't feel as 'invested' in the area and were less interested in 'getting to know people and making any contribution'. Dan also complained about the way the changes in his street affected his ability to get to know the neighbours. He said:

Everything else in the street has been turned into flats, or at least mul-tiple occupancy ... Whenever you meet new neighbours – because they're always moving – they ask, 'Have you got the upstairs or the downstairs?' Next door's *six* properties – bedsits and one-bedroom flats.

Local 'heritage', inclusion and exclusion

Exploring these broader contexts of home – the emotional impacts of degrees and scales of change and continuity – suggests a need to pay attention not just to the ways encounters with the past challenge or enhance feelings of belonging at an intimate scale, but also the social, economic and political factors which often inform such feelings. To be 'able to feel at home in a place is not just a personal matter, but also a social one', impacted by feelings of rejec-tion or not feeling welcome by the people who live in that place (Antonsich 2010: 649), and therefore 'one's personal, intimate feeling of belonging to a place should always come to terms with discourses and practices of socio-spatial inclusion/exclusion at play in that very place and which inexorably conditions one's sense of place-belongingness'. This ability to express one's own identity and be recognised as part of the local community is itself 'wed to the history of ethnic and racial relations and inflected to its core by pol-itical struggles over space and place' (Dixon and Durrheim 2004: 459). In a common move, belonging becomes 'based around a distinction between insiders and outsiders, our space and their space ... an integral part of feeling "at home" may derive from the comforting realisation of others' *absence*, as well as from a *dis-identification* with the places of others'.

This wider sense of social 'at homeness' is useful for a focus on how senses of belonging at home are enhanced or troubled by the past's imagined similarities and differences. As I have shown, these sometimes involve forms of assimilation or exclusion of 'others' in relation to, for example, decisions about who to sell one's property to. They also play out at broader scales. Whilst some rightly caution that 'not all local belonging is negotiated through the juxtaposition of incomers and incumbents' (Burrell 2016: 1603), this continues to be a key concern in terms of how far people feel included or excluded in their local neighbourhood. This might, in turn, impact on more intimate senses of home – perceived as a 'locus for and means of negotiating the dialogue between personal and communal heritage' (Robertson 2016 [2012]: 9–10). We might thus consider how different selves and others – of both the past and present – map onto broader senses of belonging. This returns to Ahmed's sense of the way some bodies are inscribed with otherness, which

> undermines the notion that cities have become a harmonious melting pot of difference. ... The notion of the stranger remains pervasive, in 'real life' and in urban theory ... 'difference' continues to be one of the most salient points of tension in many accounts of social relations, with ethnic difference presented as a particularly resilient social fault line.
>
> (Burrell 2016: 1600)

These relationships with proximate others – in particular, in relation to race, class, gender and sexuality – further inform people's narratives about the history of home. Emily and Derek described their small, distinctive line of arts and crafts houses as a remnant of the past which contrasted with their outer suburban neighbourhood, created from cheap, derelict 'infill' land. Emily related its recent construction – which meant it had 'no sense of history or depth' – with its lack of a 'sense of community', pointing to Edward Relph's idea of 'non-place' as an apt description of the suburb.[5] But tellingly, she then described how it had a history of attracting *migrant* communities, particularly from Eastern Europe. There were the interesting neighbours – a famous Hungarian ballet dancer, a woman who had worked for the Polish resistance. The house's first early-20th-century occupant had been a Polish Catholic priest which 'set the tone'. The area had been noted previously for its Italians, but recently there had been Lithuanians. The local school continued to have a 'very high proportion of immigrant children' because of the 'background of this area being hospitable'. The 'non-place' with 'no sense of community' and no sense of historical depth is, it turns out, a place where certain immigrant communities have gravitated and feel able to make home. What is implied here is that the suburb's *lack* of a sense of history allows newcomers to place their own stamp on the area more easily.

Elsewhere, in contrast, Martin remembered with a degree of regret the previous feel of their square, explaining:

> It's become gentrified basically – which I don't like ... middle class rather than working class. ... I liked it when it was edgy, you know. I liked the prostitutes. I liked the old dossers outside there and all sorts of stuff. ... One of the [prostitutes] chased my son around [laughs]. ... He had to run away from them. Anyway – *mad, mad.*

Now 'all the pubs have been done up' which, as Carol added, meant 'you lose certain charms ... the pub on the corner was a really lovely old fashioned pub' where there were 'brilliant' lock-ins. During these, Martin had talked to the old men from the council estate which formed the fourth side of the square:

> Look at the flats across there. I used to know quite a few guys over there who drank in the pub. And they said, 'You must envy us'. I said, 'What, living in a flat like that?' They said, 'Yeah. The reason being we're looking out over you whereas you're looking out over us' [laughter].

The contrast between these proximate lives – those in the council flats looking out at those living in the old houses built for the wealthy – is made light of as they shared a drink. Martin did not express embarrassment at his judgement about the difference in their homes (and by extension, lives), but entered into the ironic joke – *they* can look at his beautiful home, *he* has to look at their (presumably ugly) council block. Confused at first by the joke, he had asked with brutal honesty what he could possibly envy about 'living in a flat like that'. They poked fun at him, but with self-deprecating humour. Martin is both observer of change and a part of it, reflecting this complex positionality – nostalgic for the 'edgy' pre-gentrified times and an observer of the impact on those who had been there longer – the old men displaced, long gone, the pub now a trendy bar.

The adjacent living of people from different backgrounds in areas in transition can be more fraught. Harvey and Rochelle, for example, described not only the contrast of fortunes between their now-renovated Georgian house in relation to its previous state as a brothel and flats, but also the *continuing* disparities – the council estate at the back with its large immigrant community, a contrast highlighted through Harvey's 'borrowed view' over the estate as part of his Japanese Zen Garden. Although played down, it became clear that some of the neighbours had expressed hostility:

ROCHELLE: They are obviously people living in a much smaller space than we are. There is one very [lowers voice] badly behaved child. He's always throwing out Lego and sometimes this child aims it *directly* at us [laughs]. I mean [laughs] ... It's a targeted thing, yes ... *aiming* it at *us*.
HARVEY: He's an older teenager I think.
ROCHELLE: There was fireworks at one stage. So [pause]. I don't know. Perhaps I'm over sensitive about it. But I do feel the contrast between the people living in the flats and – *us*. At times ... I think they are just disaffected young men really. There is actually a lot of unemployment. Somebody did once throw something. We've had paint balls thrown and that glass got *broken*.

Such a contrast in lives in close proximity was noted by other participants, although sometimes the story expressed different forms of frustration. Lillian reflected on the 'unloved' adjacent council estate:

Nobody has ever liked that from the minute it was built. In fact, my friend's auntie had her little cottage demolished and moved into this block of flats. And they put her on the twelfth floor ... and she hated it so much she jumped out of the window. She died. And every time I go past there – I was very fond of her aunt – and, you know, we think it was probably she was going through the menopause, and she, you know, she was sort of fragile and everything. She was single, she didn't have – you know, suddenly her mother died and, you know ... And also on the 12th floor she could hear the lift going up and down constantly so she couldn't get any sleep. But she jumped out of the window.

The story of the friend's aunt mixed the effect of her home in the block of flats and her own emotional state and personal circumstances. Here the flat is part of the story of her demise – it both contributed to and reflected her state of mind, reinforcing an assumed vulnerability, isolation and discomfort of a woman left on her own in the wrong place. Lillian's own house, which she was anxious might eventually be demolished, was a place with a history worth preserving; the thought of the council estate being destroyed would be a relief – a stark contrast in the values given to different forms of material inheritance, the places where people choose to be and want to protect or get trapped in and need to escape.

Some participants described a reversal of their neighbourhood's fortunes or a shift in the social dynamic which had affected them at a personal level. Laura, for example, contrasted the oral testimony she had heard about the busy 'well loved' brewery community in its heyday with the transient tenants of its conversion into mostly rented flats. The looming proximity of a new major rail link directly behind the building and the noise from the ongoing building work – the subject of official complaints – combined with her sense of a lack of community in the building. She stated:

> You have all the social housing ... the immigrant communities who have come and gone – that transience. I guess in a way this place has just *adopted* that transience. And it has been quite difficult to get this place to be *home*. ... It's like a flat but not a home, despite my best attempts. I feel like I'm fighting a sort of psychogeography of the house and the area. ... You know, people have come – and *left*.

'Community' heritage

These responses to the wider context of home impact emotional senses of belonging and being 'at home' in different ways – whether the settled owners of old houses or insecure tenants in crumbling flats. The stories of social change in these inner city – mainly London – neighbourhoods reinforced negative assumptions about transience (including feelings of insecurity about the future) in contrast to a longing for stability. These assumptions appear to underpin community policies concerned with 'questions of heritage, of ownership, of discourses of past and present' as 'important elements in present-day struggles over identity and belonging, not least those related to immigration policy' (Buciek *et al.* 2006: 185). In turn, related debates about 'the politics of engagement' (Watson and Waterton 2010: 1) include concerns that people should not be seen as 'passively needing things done for them but are active subjects with values and judgements relating to heritage that hold validity in their lives' (Waterton 2005: 319).[6]

Senses of transience and tensions between different proximate community groups complicate assumptions that the practice of gaining local historical knowledge will enhance senses of connection and belonging to place, helping to create community 'cohesion'. Many participants felt that local heritage was important, but were not sure how such practices could enhance senses of belonging. Sandra, for example, felt that old buildings that had been part of the locale for a long time – like the house she lived in – had a role in creating anchors for local stability; when left in disrepair, they also acted as sirens pointing to deprivation and neglect:

> Features of places reinforce stability and continuity. ... If you bulldoze an area and put up modern stuff it starts to feel like instability and chaos. ... It is ignorant not to relate to things about this building ... that people see – the public – because of the size of it. ... It's a bit of a blot on the landscape.

Cathy had received funding to create a community photography project for local children, inspired by the famous photographer who had recorded the social life of her street from the front doorstep. She argued that although some recently-arrived communities didn't seem to 'relate to local place', community heritage – if initiated by the community *itself* – was 'incredibly important, not least for people who *arrive* in an area. Because otherwise you lose that sort of *continuity*'. But Emily in her 'no-place' suburb which 'doesn't have much of an identity' was critical of any '*forced* community regeneration', adding scathingly: 'Belonging and identity. Trying to *create* communities ... There's no *real history*. It's pastiche ... It's ephemeral'.

Brenda was also ambivalent. There were the 'long-term East Enders who are interested in the area' who want to 'flesh that [experience] out' through oral history, 'because they get stories which come down the generations and they, you know, if they've got that kind of curiosity, they think, "Well actually, you know, did that happen? I don't know"'. But when local history becomes more of a 'project', she added, it becomes about a different kind of 'knowledge': 'you've been educated in a certain way. And have got certain – *attitudes*'.

Dan, living on the main road in Hackney, east London, pointed out that local people only tended to talk to each other when they were complaining:

> If you want to create a community, go down and make a white cross on every tree in the street. And people just come out outraged because they think it's all about to be knocked down. It's a shame to always *fight* for something.

And whereas local heritage would be a useful strategy for counteracting this tendency, he mused on 'how fast things change. ... Doesn't [knowing history] just make you a little more *human*, because everyone lives in their own little world at the moment'. The problem was that cities are places 'where strangers live next door to each other', and 'people would just move away and you'd have to repeat the effort'.

Exploring and sharing the history of home as neighbourhood becomes a performative act of creating as well as reflecting value, with some participants suggesting that many heritage practices continued to be selective in deciding *who's* heritage is significant whilst others actively involved themselves as 'experts' or facilitators. Underlying participants' responses to broader contexts of heritage at home was an often anxious concern to shift from senses of transience to feelings of continuity – the need for 'anchors' in place, as Kathy Burrell suggested, outweighing senses of place as fluid, despite acknowledgement that change was inevitable and had happened, or was happening, around them. Anxieties about change, as well as divisions between perceptions of newer and more established households – were reinforced by an assumption that recent incomers would have no existing attachment to the area, and therefore would be less interested in local history – their roots and senses of 'home' remained 'elsewhere'. Witnessing, being a part of, or being impacted by local change influenced people's understandings of and values granted to the pasts of their homes, suggesting a need to acknowledge how these pasts

Figure 8.1 The end of an old pipe found new use as a hook for a dressing gown.

are imbricated in the broader contexts in which affective, embodied and emotional responses to heritage at home play out.

Considering the home in its wider local setting and social contexts suggests that a focus on domestic heritage might lend itself to insights into how more private and more public forms of engagement with the past are mediated. This leads to further questions about how heritage bodies and museums might explore what their visitors *bring* to their institutional or public spaces of their own experiences and beliefs about the past. But rather than always supporting 'people' to engage in heritage, how might heritage bodies gain insights from the ways in which people already act as their own curators, and the thoughtful insights they already bring to their understanding of how the past shapes the present and future? In turn, the book raises further questions about the ways different 'homes' at different scales continue to be configured as separate. What about, for example, the relationship between domestic homes and institutional spaces of heritage which are home *to* people, or where people have different *senses* of home, belonging or custodianship? How does this change definitions of what constitutes 'everyday' practices and experiences of heritage? And what lessons might be taken from people's experiences of their brushes with official heritage policies, where their homes are treated as 'museums' not to be tampered with, not allowed to 'evolve'?

Further questions are raised by this research. How might we be able to harness ethical and hospitable senses of custodianship and care, and, alternatively, ensure that our desire to seek connections with others does not marginalise or silence those deemed *too* different to us, or assume connections to them which *avoid* acknowledging their differences? Specifically, how can a framing of heritage in relation to home – including engaging those objects and memories one does *not* bring to a new home but await us there – impact on specific senses of belonging and home-making, such as for recent migrants? How might this be extended to consider attempts to make home within local neighbourhoods which are full of different 'others' already claiming a pre-existing sense of 'home'?

Unlike heritage sites open to the 'public', within the home encounters with the past are not usually deliberately performed or displayed – as Robertson (2016 [2012]: 2) notes, they do not 'seek to attract an audience' and therefore require careful unpacking. But fundamental to this book has been an argument that people's stories, beliefs and ideas of what is valuable about the past are equally as valid – as well as equally as contingent – as professional or 'expert' ideas and values. Just as it doesn't work to assume that 'everyday' forms of heritage are either parochial and narrowly-engaged or radically counter-hegemonic, nor does it work to assume the 'people' are a homogenous group of non-experts, to be compartmentalised, patronised or supported to understand 'the' past or 'their' pasts in ways which perform different forms of knowledge and power. Participants displayed a variety of engagements, as shown, were very conscious of their own, sometimes contradictory responses, and were often aware of broader debates about public practices of history and heritage – including the fundamental question of who gets to decide which pasts are valued and which are not. Further to this, participants interviewed for this book had themselves different levels of involvement with 'heritage' – as intelligent observers, as visitors to heritage 'sites', as professionals such as archivists, community workers, consultants, administrators, as those who have studied history or conservation in higher education and those who had published their research. The line between 'expert' and 'amateur' is thus difficult to draw. Distinguishing between them would require the impossible task of comparing degrees of knowledge and understanding by applying narrow parameters which, as described in this book, are not useful. This, in turn, returns us to this question: how do we value and utilise an engagement with embodied memories and practices and affective encounters with a home's pasts – the use of the end of an obsolete gas pipe to hang a dressing gown, or a feeling of foreboding on the turn of a staircase – in relation to broader, critical engagements with 'the past' in wider context? If we believe that both need to be acknowledged as valid – as explored in this chapter in relation to the wider local scales of heritage at home – how could they be made to speak to each other without assuming they are in opposition, or precluding one in favour of the other, or subsuming one *within* the other? What kind of colonising of knowledge gets reiterated, or what new forms of knowledge might get created? Taking seriously the ways that people's different emotional and embodied experiences and encounters with the home's various pasts are bound up not just with their own memories, experiences and histories but also with the pasts and futures of others, as has been shown, allows new insights to emerge into the beliefs, ideas and values underpinning people's relationships to a variety of selves and others. Fundamentally, as this book shows, people will continue to engage with heritage in their homes in many different, sometimes idiosyncratic, ways. If we too quickly impose narrow ideas about what we *should* value and how we *should* engage with such forms of heritage, are we not just trying to maintain our own distancing boundaries, contain what feels unruly? As Wallace Stevens wrote in his poem, 'Notes Toward a Supreme Fiction':

> *The Sun*
> *Must bear no name, gold flourisher, but be*
> *In the difficulty of what it is to be*

Notes

1 The book is populated by many 'others' whose voices are not directly heard. The different ways that participants speak for past lives and for the lives of more recent residents or neighbours is, for example, matched by the book's limitation in only being able to speak for some lives and not others.

2 Some commentators also rather disparagingly suggest that an increased interest in the past is fuelled by a baby-boomer generation with the resources to indulge in

history as a pastime – an 'aging population affluent and literate enough to take an unprecedented interest in family and local history ... guarantee[ing] history a key audience base' (Dresser 2010: 43).

3 Hibbing *et al.* (2017: 234) also describe how 'political views will influence how [people] perceive the past', suggesting that 'people differ in their views toward the past based on ideology'.

4 The *Right to Buy* scheme was introduced in the UK's Housing Act 1980 by the Conservative Government, based on a belief in encouraging self-sufficiency and an assumption that everyone aspired to own their home. Council properties, later extended to housing associations, were offered for sale to tenants at discount prices, although rules have since tightened.

5 Relph's belief that the 'weakening of distinct and diverse experiences and identities of places – is now a dominant force' (Relph 1976: 45) formed his distinction between place and 'placelessness'. He pointed to the 'persistent sameness and unity which allows that [place] to be differentiated from others' and the way 'individuals practice a multiplicity of places: they are sometimes insiders and sometimes outsiders, following the social context of their action and their intentions' (Seamon and Sowers 2008: 616).

6 Waterton and Smith (2010) point out that ideas of 'community' within the field of heritage are replete with tensions between different groups and the way 'memory, place, identity and cultural expression' are defined and negotiated may differ. There has been a recent shift to consider 'community' in terms of relational encounters between affective bodies (Walkerdine 2016), and to everyday encounters in general, including the role of 'conviviality' in creating sociality. Kathy Burrell (2016: 1600) acknowledges warnings about the 'danger of almost essentialising difference itself in the study of social relations, and of *flattening* the range of social encounters identified, making fewer distinctions between passing smiles in coffee shops between strangers and more prolonged or deliberate social interactions'. However, she argues that the 'move towards encounter and conviviality ... has been particularly valuable for the more explicit groundedness it offers in analysing and locating social interactions'. This includes recognising the role of short-lived or transient encounters in creating social relations – akin to previous ideas of 'weak ties' from episodic everyday encounters and 'everyday life micro-publics' (Valentine 2008; Studdert and Walkerdine 2017; Neal *et al.* 2019). But these may not be sufficient to counteract an absence of feelings of belonging to place, leading to senses of loneliness, isolation, alienation, and dis-placement (Antonsich 2010: 649).

References

Ahmed, S (2004): 'Affective Economies'. *Social Text* 79 22.2: 117–139.

Antonsich, M (2010): 'Searching for Belonging – An Analytical Framework'. *Geography Compass* 4.6: 644–659.

Bachelard, G (1994 [1958]): *The Poetics of Space*. Boston: Beacon Press.

Bennett, G (1999): *Alas, Poor Ghost! Traditions of Belief in Story and Discourse*. Logan: Utah State University Press.

Bennett, G (1987): *Traditions of Belief: Women, Folklore and the Supernatural Today*. London: Penguin Books.

Bennett, G and Bennett, K (2000): 'The Presence of the Dead: An Empirical Study'. *Mortality* 5.2: 139–157.

Bennett, J (2018): 'Narrating Family Histories: Negotiating Identity and Belonging through Tropes of Nostalgia and Authenticity'. *Current Sociology* 66.3: 449–465.

Blunt, A (2003): 'Collective Memory and Productive Nostalgia: Anglo-Indian Homemaking at McCluskieganj'. *Environment and Planning D: Society and Space* 21: 717–738.

Blunt, A and Sheringham, O (2019): 'Home-City Geographies: Urban Dwelling and Mobility'. *Progress in Human Geography* 43.5: 815–834.

Bonnett, A (2016): *The Geography of Nostalgia: Global and Local Perspectives on Modernity and Loss.* Abingdon: Routledge.

Boym, S (2001): *The Future of Nostalgia.* New York: Basic Books.

Buciek, K, Bærenholdt, J, and Juul, K (2006): 'Whose Heritage? Immigration and Place Narratives in Denmark'. *Geografiska Annaler: Series B, Human Geography* 88.2: 185–197.

Buckle, C (2017): 'Residential Mobility and Moving Home'. *Geography Compass* 11.5: 1–11.

Burrell, K (2016): 'Lost in the "Churn"? Locating Neighbourliness in a Transient Neighbourhood'. *Environment and Planning A* 48.8: 1599–1616.

Costas, J (2013): 'Problematizing Mobility: A Metaphor of Stickiness, Non-Places and the Kinetic Elite'. *Organization Studies* 34.10: 1467–1485.

Cresswell, T (2010): 'Towards a Politics of Mobility'. *Environment and Planning D: Society and Space* 28: 17–31.

Cresswell, T (2012): 'Value, Gleaning and the Archive at Maxwell Street, Chicago'. *Transactions of the Institute of British Geographers* 37.1: 164–176.

Crouch, D (2015): 'Affect, Heritage, Feeling'. In E Waterton and S Watson (eds.) *The Palgrave Handbook of Contemporary Heritage Research.* London: Palgrave Macmillan: 177–190.

De Certeau, M (1984): *The Practice of Everyday Life.* Berkeley: University of California Press.

Dixon, J and Durrheim, K (2004): 'Dislocating Identity: Desegregation and the Transformation of Place'. *Journal of Environmental Psychology* 24.4: 455–473.

Dresser, M (2010): 'Politics, Populism, and Professionalism: Reflections on the Role of the Academic Historian in the Production of Public History'. *Public Historian* 32.3: 39–63.

Hibbing, M, Hayes, M, and Deol, R (2017): 'Nostalgia Isn't What It Used to Be: Partisan Polarization in Views on the Past' *Social Science Quarterly* 98.1: 230–243.

Jones, S (2010): 'Negotiating Authentic Objects and Authentic Selves: Beyond the Deconstruction of Authenticity'. *Journal of Material Culture* 15. 2: 181–203.

Landsberg, A (2004): *Prosthetic Memory: The Transformation of American Remembrance in the Age of Mass Culture.* New York: Columbia University Press.

Ley, D (2003): 'Forgetting Postmodernism? Recuperating a Social History of Local Knowledge'. In *Progress in Human Geography* 27:5, 537–560.

Lorimer, H (2013): 'Scaring Crows'. *Geographical Review* 103.2: 177–189.

Hamilton, P and Ashton, P (2003) 'At Home in the Past: Initial Findings from the Survey', *Australian Cultural History*, 22: 5–30.

Massey, D (1991): 'A Global Sense of Place'. *Marxism Today* June 1991: 24–29.

Massey, D (1995): 'Places and Their Pasts'. *History Workshop Journal* 39: 182–192.

May, V (2017): 'Belonging from Afar: Nostalgia, Time and Memory'. *Sociological Review* 65.2: 401–415.

Miller, D (2001): 'Behind Closed Doors'. In Miller, D (ed) *Home Possessions: Material Culture Behind Closed Doors*, 1–19. Oxford: Berg.

Morley, D (2000): *Home Territories: Media, Mobility and Identity*. London: Routledge.

Moussouri, T and Vomvyla, E (2015): 'Conversations about Home, Community and Identity'. *Archaeology International* 18: 97–112.

Neal, S, Bennett, K, Cochrane, A, and Mohan, G (2019): 'Community *and* Conviviality? Informal Social Life in Multicultural Places'. *Sociology* 53.1: 69–86.

Owens, A, Jeffries, N, Featherby, R, and Wehner K (2010): *From the Unusual to the Banal: The Archaeology of Everyday Life in Victorian London*. Museum of London Archaeology Research Matters, No. 4. London: Museum of London.

Relph, E (1976): *Place and Placelessness*. London: Pion Limited.

Robertson, I (ed.) (2016 [2012]): *Heritage from Below*. New York: Routledge.

Robertson, I (2015): 'Hardscrabble Heritage: The Ruined Blackhouse and Crofting Landscape as Heritage from Below'. *Landscape Research* 40.8: 993–1009.

Seamon, D and Sowers, J (2008): 'Place and Placelessness, Edward Relph'. In P Hubbard, R Kitchen, and G Valentine (eds.): *Key Texts in Human Geography*. London: Sage: 43–51.

Strangleman, T (2013): ' "Smokestack Nostalgia," "Ruin Porn" or Working-Class Obituary: The Role and Meaning of Deindustrial Representation'. *International Labor and Working-Class History* 84: 23–37.

Studdert, D and Walkerdine, V (2017): 'Being in Community: Re-Visioning Sociology'. *Sociological Review* 64.4: 613–621.

Valentine, G (2008): 'Living with Difference: Reflections on Geographies of Encounter'. *Progress in Human Geography* 32.3: 323–337.

Walkerdine, V (2016): 'Affective History, Working-Class Communities and Self-Determination'. *Sociological Review* 64.4 699–714.

Waterton, E (2005): 'Whose Sense of Place? Reconciling Archaeological Perspectives with Community Values: Cultural Landscapes in England'. *International Journal of Heritage Studies* 11.4: 309–325.

Waterton, E and Watson, S (eds.) (2010): *Culture, Heritage and Representation: Perspectives on Visuality and the Past*. Farnham, Surrey: Ashgate Publishing.

Waterton, E and Smith, L (2010): 'The Recognition and Misrecognition of Community Heritage'. *International Journal of Heritage Studies* 16.1/2: 4–15.

Watson, S (2018): 'Emotional Engagement in Heritage Sites and Museums: Ghosts of the Past and Imagination in the Present'. In S Watson, A Barnes and K Bunning (eds.) *A Museum Studies Approach to Heritage*. Abingdon: Routledge.

Woodham, A, King, L, Gloyn, L, Crewe, V, and Blair, F (2017): 'We Are What We Keep: The "Family Archive", Identity and Public/Private Heritage'. *Heritage & Society* 10.3: 203–220.

Wylie, J (2012): 'Dwelling and Displacement: Tim Robinson and the Questions of Landscape'. *Cultural Geographies* 19.3: 365–383.

Young, I M (2005): *On Female Body Experience: 'Throwing Like a Girl' and Other Essays*. Oxford: Oxford University Press.

Index

accidental 49, 101, 156, 158, 159, 162
aesthetics 1, 11, 94, 128, 136, 139, 154,
  163, 164, 170, 174, 175, 177, 181, 182,
  187–189
affective experience 1, 3, 85, 168;
  atmosphere 75, 200; bodies 222n6;
  economies 15, 24, 196n2, 199, 201, 24;
  encounters 30n7, 167, 198, 206, 221;
  forms of knowing 73; heritage 56, 220;
  inheritance 15; memory 21; politics
  15; relationships 15, 19, 22, 125, 164,
  208; responses 11, 29n6
agency 7, 13, 18, 24, 25, 86, 164, 167,
  171, 173, 183, 201, 203
Ahmed, S 1, 14, 15, 30n7, 30n9, 31, 39,
  77, 86, 143, 144, 196n2, 196, 199, 209,
  215, 222
Al-Ali, N (with Koser, K) 5, 6, 31
amateur 4, 12, 67, 221
Anderson, B 30n7, 30n9, 31n11, 31
Antonsich, M 208, 214, 222
anxiety 92, 113, 133, 168, 210,
  217, 219
approval/disapproval 108, 129, 131, 135,
  176, 191
archaeological dig 125, 154, 158,
  168n1, 189
archaeology 22, 25, 26, 29n3, 29n6, 48,
  49, 50, 149, 153, 158, 167, 181, 189
architecture 7, 50, 64, 92, 108, 122, 159,
  177, 178, 180, 188, 196n3
archive 9, 10, 19, 21, 26, 45, 47, 70,
  147–149, 152, 162, 164, 167, 198
Armitage, S 147, 149
art deco 120, 186
Ashton, P 3, 11, 18, 19, 31, 205
Atkinson, D 3, 4, 22, 32
atmosphere 1, 11, 15, 23, 24, 30n9,
  49, 56, 73, 75–81, 83, 98, 109, 122,
  137, 164, 172, 202, 203, 195, 196n3,
  205, 206
attic 25, 72, 134, 147, 157, 163, 165

Bachelard, G 5, 8, 147, 200
Backe-Hansen, M 13
banister 94, 95, 113, 181, 185
Barclay, K 21
Barratt, N 13
bedroom 45, 53, 57, 68, 69, 81, 84, 96,
  126, 127, 147, 149–152, 155, 186,
  193, 212, 214; see also basement;
  bathroom; kitchen; living room
basement 107, 142, 182, 183, 210
bathroom 138, 178, 182
belief 2, 12, 14, 19, 22–26, 28, 28n1, 29,
  48, 62, 66, 68, 71, 75, 76, 81, 83, 85,
  86, 91, 92, 97, 103, 113, 133, 137–139,
  152, 164, 165, 167, 170, 174–176, 181,
  190, 192, 195, 199, 201, 202, 220, 221,
  222n4, 222n5
Bell, K 23, 75, 85
belonging 1, 2, 5, 6, 8, 9, 11, 13–16, 29,
  86, 91, 93, 99, 102, 109, 110, 114, 115,
  117, 119, 121, 123, 124, 127, 131, 133,
  134, 141, 143, 160, 162, 164, 165, 167,
  177, 193, 198–201, 203–211, 213–215,
  217, 218, 220, 222
Berger, J. 17, 20, 29n5, 48, 66
biography 47, 66, 148, 160, 196n1
Blakely, H (with Moles, K) 9, 123, 167
blessing 117, 118, 123, 124, 143, 211;
  see also curse
Blunt, A 29, 205
bodies 6, 14–16, 24, 28n1, 30n7, 177,
  215, 220, 222n6
Bonnett, A 205
Burrell, K 25, 28, 205, 208, 209, 215,
  219, 222

care 12, 17, 66, 98, 114, 118, 139–141, 143, 155, 157, 159, 168, 171, 173, 174, 187–189, 200, 212, 213, 220
caretaker 172, 194
census: children 68, 77, 98; data 61, 77, 92; deaths 70, 99; length of stay 110, 210; names 60, 61, 64, 75, 76; number of residents 104; records 46, 68, 70, 75, 98, 103, 139, 193; research 27
change: attitude to 10, 177, 210; historical 29n5; home, material 13, 52, 69, 76, 80, 96, 108, 124, 125, 126, 129, 132, 135, 136, 160; home, relationship to 50, 81; human nature 61, 97; inhibition about/resistance to 174, 175, 177, 201, 205, 219; judging previous 195; justifying 163, 188; local 28, 45, 78, 104, 130, 204, 210, 212; negotiating continuity and change 203, 204, 209, 214; previous, to home 178, 179, 181, 184; socio-economic 104, 138, 206, 213, 217; styles/fashions 175, 179, 185, 188; *see also* continuity
children 27, 52, 58, 61, 68, 77, 78, 97–99, 104–106, 108, 130, 138, 140, 148, 150, 151, 187, 215, 218; *see also* census
coincidence 64, 99, 100, 107, 114, 127, 137
Collingwood, R G 21
colour 94, 126, 150, 176, 181, 188
comfort: as bodily dis/comfort 57, 105, 108, 114, 138, 179, 180, 181, 200, 213; in comfortable place 70, 97, 195, 196n3, 211; as comforting routines 5, 141, 207, 214; feelings of 68, 69, 77, 83, 93, 111, 102, 103, 112, 131, 182, 191, 217
community heritage 217
connecting: abstractly, to home's temporalities 93; to earlier residents 95; emotionally, to home 12, 13, 51, 91, 92, 94, 99; imaginatively 60; to future others 162; to past others 16, 21, 91, 103, 113, 204; *see also* coincidence; continuity
continuity 6, 45, 49, 69, 93–96, 113, 114, 128, 147, 149, 198, 203–205, 208, 209, 213, 214, 218, 219; *see also* change
convention 15, 17, 19, 20, 30n9, 67, 144n1, 167, 187, 191, 201
craftsman/ship 155, 173, 174, 184, 206
Cresswell, T 5, 10, 198, 207
Crouch, D 3, 30n9, 198

curse 78, 79, 117, 118, 140, 143, 201; *see also* blessing
custodian/ship 128, 143, 164, 170, 172, 173, 175, 179, 188, 190, 194, 195, 200, 213, 220

dark 25, 45, 56, 70, 72, 73, 76, 83, 100, 107–109, 192
dead 12, 13, 21, 48, 73, 80, 93, 99, 101, 106, 115n1, 117, 152, 154, 162, 168n1, 171
death 29, 45, 64, 70, 81–83, 97, 102, 105, 111, 124, 140, 141, 166, 168, 211, 212; *see also* census
De Certeau, M 3, 8, 200
décor 11, 16, 62, 71, 118, 119, 127, 171, 174
decoration 55, 117, 118, 120
design 10, 53, 108, 128, 173, 174, 175, 176, 178, 181, 189, 195, 206
DeSilvey, C 9, 25, 94, 144n1
Digby, S 9, 25, 94, 144n1
digging: in garden 49, 153; up cat's remains 168n1; up flowers 125; up history 153, 158
display 4, 10, 148, 163, 167, 182, 185, 191–194, 208, 220
doors 12, 52, 58, 94–96, 105, 108, 150, 155, 163, 175, 187
Dragojlovic, A 8, 16, 30
Dresser, M 20, 222n2

Edensor, T 5, 8, 9, 24
Edwardian 47, 53, 55, 70, 107, 173, 175, 177, 178, 186, 190, 212
embodied: contact 94; encounters 2, 131, 164, 198, 202; heritage 22, 23; knowledge 29n6, 45, 49, 50, 73; processes 1, 3, 5, 15; responses to past others 91, 96, 220; tradition 161
emotional: connection 91, 92, 114, 125, 133, 143, 102, 168n1, 208, 217; definitions of 28n1; encounters 2, 138; experience of heritage 18, 66, 220; experience of the past at home 22, 50, 67, 76, 82, 85, 94, 149, 167; impacts 5, 17; meaning 45, 148; responses to others 14, 58, 77, 114, 157; responses to past homes 126, 127, 132, 111; responses to research 11, 19, 21, 48, 113; temperature 5, 17; *see also* feelings
empathetic 21, 129, 168

empathy 16, 17, 22, 28, 29, 54, 114, 201
enchantment 7, 25, 29n6, 31n11, 15
energy 151, 79–82, 103
era 30n9, 47, 59, 60, 71, 98, 105, 108, 120, 133, 163, 174, 176, 178–180, 183, 185, 187, 188, 195, 196n2, 206, 213
ether 80, 81, 82, 85
ethereal 73, 82
etiquette 112, 120, 128, 131, 132
everyday: encounters 4, 167, 210, 222n6; experience 96; heritage 2, 3, 66, 201, 220; homemaking 1, 200; life 4, 55, 58, 147, 192, 198, 209; routines 200; spaces 8, 20, 86, 143, 201
exchange: between selves and others 202; of contracts 111; of experience 141; of gifts 123; of information 143; of ownership 117, 124; of sense of home 137; process 117, 118, 120, 142, 143
exorcise 81, 121; *see also* blessing; curse; ritual
experiences 1, 14, 28n1, 221
expert 2, 4, 19, 22, 48, 56, 66, 165, 189, 190, 203, 219–221
expert-amateur binary 67
expertise 20, 47

familiar 11, 16, 25, 26, 75, 97, 94, 165, 209; cultural categories 28n1; emotions 8; homemaking practices 6, 27, 155, 187, 200; knowledge 70, 205; memory 8, 16, 200, 203, 221; routine 16, 58, 66, 204; sense of familiarity 11, 125, 137, 200, 204; spaces 91; strange within familiar 202; *see also* everyday, ritual
family 6, 97; beyond kinship 114; borrowed 100; connection to place 102; extended 13, 100, 101; of future residents 138; happy 98, 99, 108; history 16, 19, 45, 46, 99, 101, 110; home 64, 77, 80, 83, 98, 119, 135, 147, 190, 191; legacies 201; lore 147; memories 102, 125; objects 148, 162, 163; of past residents 127, 129, 140, 141; research 55, 99, 165, 205, 222n2; rituals 165; of servants 108; stories 10, 16; tree 13
feelings 2, 23, 28n1, 30, 70, 91; about heritage 24, 141; about past homes 126, 127; about past residents 72, 73, 97, 157; of belonging 1, 8, 86, 99, 114, 127, 134, 143, 208, 211, 213, 214;

of continuity/permanence 93, 95, 113, 219; of fascination/awe 49, 93; of guilt 182, 183, 184, 200; negative 11, 15, 29n6, 68, 85, 121, 206; of past residents 55, 61; of/in place 8, 77, 78, 82, 83, 91, 126, 137, 159, 171, 176, 179, 192; of privilege 103, 172; sharing 142; structure of feeling 22; of transience 155, 214; *see also* emotional
Ferber, M 26
Finn, C 26
fireplace 185
freeholder 70, 79, 129, 189
Freud, S 9, 25, 29n3
future: anxieties for 79, 136; generations 78; heritage 29n3, 144n1; imagined 17, 117; of the home 114, 117, 133; residents 2, 28, 91, 92, 97, 114, 133

garden 36, 49, 51, 52, 57, 59, 80, 95, 106, 108, 112, 122, 125, 129, 131, 134, 135, 148, 150, 152–154, 163, 167, 168, 181–184, 193, 204, 213, 216; *see also* plants
Gelder, K (with Jacobs, J) 26
Gell, A 24
gender 3, 5, 14, 15, 27, 30n8, 66, 76, 77, 85, 97, 98, 196n3, 215; *see also* women
genealogy 12, 13, 141
generic: aspects of past residents 58, 63, 69, 111, 112; engagement with past residents 59; facts/knowledge 60, 114; events 97; roles 97
Georgian 47, 51, 60, 62, 71, 92, 98, 104, 131, 137, 174, 177, 178, 180–183, 185, 186, 188, 216; *see also* modern; Victorian
ghosts 14, 22, 23, 52, 54, 57, 62, 68–70, 72–78, 81, 83, 85, 98, 101; *see also* haunting; more-than rational; uncanny
gift 118, 123, 124, 134, 154, 163, 166, 201
Gordon, A 22
gravestone 111, 148, 151, 170, 192
Gregson, N (with Metcalfe, A and Crewe, L) 6, 9
Grossman, A 9
guilt 99, 103, 104, 114, 123, 182, 184, 200

happenchance 158
haptic 24, 51, 93, 94, 114, 200; *see also* proximity; touch
Harries, J 24, 164

Harrison, R 5, 29n3, 143n1, 144n1
Harvey, K 5, 18, 20, 21, 66
haunting 16, 79; *see also* ghost; more-
    than-rational; uncanny
Heimlich 9, 76, 162; *see also Unheimlich*
Hepworth, K 21
heritage 1–6, 8, 10, 11, 12, 15, 18, 19, 20,
    22–24, 29n3, 29n6, 30n9, 48, 56, 65,
    66, 86, 117, 143n1, 144, 167, 179,
    189–194, 196n2, 196n2, 198, 200–204,
    206, 214, 215, 217–222n6
heteronormative 98, 111, 115
Hewison, R 18
hidden 9, 10, 24, 25, 29n3, 49, 50, 52,
    76, 110, 111, 134, 144, 148, 149, 154,
    155, 157, 160, 162–164, 167, 192, 194,
    202, 203
Hill, J 10, 164
Hirsch, M 16
Hoggard, B 164
homemaking 1, 6, 7, 9, 28, 91, 143, 187
Hoskins, A 16
hospitality 70, 131, 141, 143, 200
house history 12, 26, 46, 86, 101,
    130, 133
Hurdley, R 6, 196n1

identity 3, 4, 8, 10, 15, 26, 28–30, 76, 92,
    99, 111–114, 160, 164, 167, 171, 175,
    178–180, 183, 185, 189, 195, 199, 200,
    205, 206–208, 214, 218, 222n6
imagination 1, 7, 15, 17, 19, 21, 22, 24,
    26, 28, 31n12, 52, 57, 62, 63, 65, 66,
    68, 69, 74, 75, 77, 85, 91, 110, 125, 140
inheritance 2–4, 15, 46, 50, 51, 64, 117,
    120, 135, 140, 143, 150, 193, 201, 202;
    affective 15; domestic 1; family 182;
    inadvertent 157; material 111, 138,
    159, 175, 187, 200, 217; natural 152;
    objects 143, 147, 148, 149, 154, 155,
    157, 165, 166, 170, 171, 173, 186, 195,
    199; passing on 9, 128, 133, 134, 140,
    162, 200; wall 152
in keeping 137, 176, 177, 180–182,
    195
integrity 31n12, 175, 176, 178, 180,
    188, 195

Jenkins, K 31n12
Jentsch, E 25
Jones, S 23, 24, 86, 202
judgement 119, 120, 125, 139, 174, 176,
    186, 205, 216, 218

Kean, H 20, 131, 168n1
kinship 14, 114
Kirshenblatt-Gimblett, B 23, 86
kitchens 53, 57, 77, 97, 107, 108, 119,
    120, 129, 136, 147, 151, 154, 156, 163,
    176, 181, 182, 196n3
Kitson, J (with McHugh, K) 7, 10, 22
knowledge: difficult 29n6; historical 17,
    31n12, 165, 174, 177, 190, 218, 221;
    knowledge-making 26; local 58, 92,
    204, 218; of home 45, 48, 50, 75, 76,
    84, 134, 135, 191; of past residents
    13, 61, 69, 210; *see also* embodied;
    expertise; generic; imagination;
    research

LaCapra, D 31
Ladino, J 15
Landsberg, A 16, 17, 19, 31n11, 199
Lang, R 11, 16
Lawrence-Zuniga, D 7, 29n6
leaving: home 79, 80, 133, 136, 143,
    156; mark for future 159, 160, 199;
    memories 8; objects behind 120,
    158, 167
lesbian 111–113, 115n2
Liddington, J 19, 115n1
light 54, 62, 64, 107–109, 111, 114, 119,
    125, 135, 153, 174, 178, 181, 184, 186;
    see also dark
Lipman, C 14–16, 25, 26, 27, 30, 75, 76,
    85, 162
Lipman, C (with Nash, C) 13, 114
living rooms 52–54, 70–72, 108, 120, 147,
    156, 176, 186, 187, 192–194
local: belonging 206, 214; history 26, 45,
    47, 92, 193, 204, 205, 218, 219, 222n2
Lorimer, H 2, 207
Love, H 113
Lowenthal, D 18
Lyon, D 6

MacDonald, F 29, 3
Maddrell, A (with Sidaway, J) 168n1
maintenance 4, 48, 79, 158, 195, 205,
    212, 213
Massey, D 5, 167, 206, 207
material culture 5, 9, 24, 25, 27, 28, 193
materiality 11, 24, 52, 80, 115n2, 174
May, V 8
McFarlane, C 30n7
memorials 141, 148, 160, 168, 170,
    191–193, 196n2

memory 1, 3, 5, 7–9, 12, 15, 16, 18, 19, 21, 22, 24, 30n7, 46, 77, 81, 127, 131, 133, 141, 147, 149, 181, 193, 200, 222n6
middle class 7, 83, 104, 210, 212, 215
Miller, D 5, 11, 12, 94, 114, 115, 200
modern: comforts 138, 181; conception of heritage 144n1, 208; fixtures/furniture 120, 163, 176–178, 180, 182, 195; homes 78, 98, 159, 174; modernise 68, 76, 111, 138, 143, 176, 180; modernising 1950s and 1960s 157, 173, 177, 180, 186, 188, 206, 212, 218; post-modern 178; pre-modern 10, 174; rooms 185; technologies 51, 180
modernisation 143, 177
more-than rational 2, 17, 68, 85, 86, 199, 202; *see also* ghost; haunting; uncanny
Morley, D 14, 210
Morphy, H 24
Moussouri, T (with Vomvyla, E) 4, 198
mundane 3–5, 86, 203; *see also* everyday
Munslow, A 31n12
murder 81–83
museum 3, 4, 10, 17, 26, 27, 29n5, 115n1, 147, 149, 173, 178–181, 190–192, 220
Myerson, J 29n5
myth 20, 22, 64

names 54, 56, 60, 61, 77, 148, 159, 160, 167
Nash, C 5, 13, 23
neighbourhood 8, 62, 78, 97, 104, 131, 132, 137, 138, 203–205, 208–210, 212, 214, 215, 217, 219, 220
Nora, P 3
nostalgia 18, 85, 106, 110, 138, 174, 182, 205–207

objects 1, 6, 9–12, 16–18, 23, 24, 26, 28n2, 30
official 3, 4, 10, 18, 23, 24, 65, 86, 189–191, 194, 196n2, 201, 202, 217, 220
Oram, A 115n2
ordinary 3, 4, 12, 46, 55, 56, 63, 86, 95, 190, 193, 203
original 7, 20, 25, 50, 53, 54, 60, 92–95, 103, 113, 120, 126, 127, 133, 135, 138, 139, 148–150, 153, 157, 159, 162, 163, 165, 171, 173–181, 183–187, 193–196, 200, 201, 208, 213; aesthetic 139, 181, 203; architecture/design 92, 176, 177,

181; banister 113; craftsmanship 173, 174, 184; door 54, 150, 181, 194; era 133, 163, 176, 179, 185; features and fixtures 93, 94, 103, 127, 149, 174–176, 185, 187, 213; fireplace 50, 159, 173, 175, 181, 186; floorboards 162; grate 53; homeland 138; identity 183; instructions 150; integrity 24, 175, 176, 178, 180, 189, 195; intention 175; layout 180, 183; meaning 148; objects 95, 176, 178, 200, 208; occupants 60; owner 25, 127, 135, 180, 201; staircase 53, 95; trees 153; use 139, 165; wallpaper 157; window 120, 126, 157, 171, 176
otherness 114, 115, 115n2, 199, 202, 206, 208, 215
Owens, A (with Jeffries, N, Featherby, R and Wehner, K) 153, 201
ownership 5, 12, 13, 110, 117, 122, 133, 141, 218

palimpsest 178, 179, 184
pastiche 150, 176–179, 218
Pearce, S 29n2
permanence 93, 95, 102, 113; *see also* continuity; transience
Perry, S 29n6
photographs 10, 53, 59, 71, 96, 100–102, 105, 106, 127, 134, 141, 147, 148, 150, 154, 165, 188, 191, 193, 218
Pihlainen, K 21, 31n12
Pile, S 26
plants 11, 123, 125, 126, 154, 181
presence: continuing 16, 121, 124, 134, 143; material 1, 12, 25, 53, 142, 150, 167, 198, 200, 201; of others 13, 25, 45, 91, 98, 101, 111, 157, 199, 202, 208; of the home's past 11, 15, 22, 23, 68, 113, 117, 164, 170, 195
preservation 7, 9, 22, 56, 144, 156, 158, 189
previous residents 2, 47, 52, 57, 60, 64, 68, 69, 74, 95, 103, 109, 113, 118, 119, 122, 125, 127, 129, 139, 143, 147, 151, 157, 159, 165, 181, 185, 187, 208
private: as domestic interior 5, 9, 11, 14, 15, 45, 91; display 193; encounters 16, 17, 132, 171, 220; events 68, 148; garden 193; histories 18, 19, 190; meaning making 24, 202; memorial 170, 192; memory 8; private spaces as shared 2, 13, 199; sense of belonging

8; spaces within home 69, 83, 185, 194, 203; tenants/housing 27, 131, 195, 211, 213; *see also* public
professional: 4, 18, 20, 47, 48, 56, 160, 167, 212, 220, 221; attitude 56; distance 19, 22, 48, 66; heritage 4; instinct 160; knowledge 47, 48; practice 167, 221; training 20; values 220; *see also* experts
progressive 6, 179, 195, 207
proximity 14, 25, 204, 213, 216, 217
public: display 192, 182, 185; heritage 48, 190–192, 221; history 17–20, 56, 190, 205, 115n2, 205; housing 211; memorial 191; memories 17; space 5, 9, 16, 26, 45, 194, 198, 204, 220; the public 21, 22, 190, 220; *see also* local heritage; private

relatedness 1, 2, 13, 96, 114, 141, 167, 192, 201
Relph, E 215, 222n5
renovation 14, 46, 48, 50, 52, 103, 124, 135, 143, 148, 149, 154, 170, 174, 177, 180, 185, 187, 200
reproduction 23, 176, 177, 186
research 2, 4, 6, 9, 11–14, 19–21, 26–29n4, 45–49, 52, 54, 55, 59, 66, 72, 75, 81, 83, 96, 99, 101, 115n1, 133, 165, 175, 177, 180, 181, 198, 199, 204, 205, 208, 209, 211, 228, 221
responsibility 7, 17, 20, 159, 172, 173, 175, 183, 194, 200
Right to Buy 211–222n4
rituals 8, 45, 121, 122, 124, 133, 134, 142, 143, 159–161, 202; of gift exchange 123; of making home 200, 201; of reburial 162, 164–167; of transfer 118
Robertson, I 4, 9, 18, 55, 171, 215, 220
rooms 11, 49, 52, 53, 56, 57, 80, 96, 99, 104, 105, 108, 109, 130, 138, 147, 162, 163, 177, 181, 185, 186, 191, 194
Rose, M 25, 86
Rosenzweig, R (with Thelen, D) 18
routines 2, 5, 8, 16, 57, 58, 85, 96, 114, 200, 204
rural 27, 56, 102, 106, 113

salvage 28, 29n3, 64, 120, 122, 137, 144, 156, 158, 176
Samuel, R 3, 18, 19, 22

Sather-Wagstaff, J 8, 19, 29n6
Schofield, J 20, 29n6, 67
scrapbooks 10, 47, 56, 134, 150, 163
self/other relationship 117, 143, 198
sentimentality 48, 66, 141, 151, 205, 207
servants 53, 107, 108, 111, 150, 193, 196n3, 199, 212
sexuality 3, 5, 112, 114, 139, 215
Shanks, M 25
Shuman, A 17, 55
Smith, L 3, 18, 23
Smith, R 21
social context 10, 15, 16, 26, 98, 149, 165, 200, 201, 220, 222n5
spectral heritage 23, 202
spiritual 23, 26, 72, 73, 78, 85, 91, 202
staircase 12, 49, 52, 53, 74, 76, 95, 107, 158, 159, 185, 221
Strangleman, T 19, 29n2, 205
Strathern, M 14
streets 12, 52, 70, 81, 84, 93, 96, 99, 104, 121, 131–133, 138, 141, 153, 176, 184, 204, 209, 210, 214, 218
storage 151, 157, 164, 181
stories 1, 10, 13, 17, 23, 24, 28, 28n2, 29n2, 29n5, 46, 50, 55, 59, 60, 63, 64, 69–71, 102, 122, 130, 133, 142, 143, 149, 153, 154, 156, 163, 193, 202, 217, 218, 220
Studdert, D (with Walkerdine, V) 8, 30n8, 222n6
suburb 7, 27, 45, 46, 47, 48, 56–58, 62, 72, 78, 82, 117, 127, 165, 215, 218
Sumartojo, S (with Pink, S) 30n9

taste 119, 139, 163, 177, 185, 187, 188, 195, 200
temporary 92, 133, 172, 194, 210; *see also* transience
tenant 27, 78, 120, 137, 156, 195, 209, 211–214, 217, 222n4
Tindall, G 12, 14
Tolia-Kelly, D 6, 8, 15, 22, 29n6, 30n7, 143
touch 11, 17, 24, 48, 54, 94, 95, 160, 164, 200, 220; *see also* haptic; proximity
traces 1, 9, 10, 22, 25, 49, 52, 54, 66, 80, 81, 112, 117, 150, 155, 158, 179, 198
tradition 5, 7, 8, 13, 18, 22–24, 28n1, 85, 123, 148, 151, 160–162, 164, 165, 167, 174, 198, 201, 202, 208
transience 11, 110, 155, 208, 210, 211, 214, 217–219

trees 13, 52, 70, 93, 94, 151, 153, 182, 204, 213
Tuan, Y-F 5

uncanny 9, 14, 15, 25, 26, 30, 60, 75, 76, 97, 131, 198, 202; *see also Unheimlich*
unhappy 30n9, 77, 83
Unheimlich 9, 25, 76; *see also Heimlich*
Uprichard, E 30
urban 6, 8, 27, 94, 210, 215

Valentine, G 143, 222n6
value 219, 221, 7, 10, 14, 17–19, 22, 23, 55, 66, 93, 95, 109, 114, 141, 147–149, 154, 159, 162–164, 168, 173, 174, 182, 184, 185, 187, 194, 195, 201, 210
valued 18, 86, 153, 167, 175, 198, 221
values 1, 2, 4, 7, 10, 91, 103, 113, 133, 136, 137, 154, 164, 167, 173, 178, 181, 183, 185, 188–190, 192, 195, 196, 199, 201, 206, 207, 217–221
Varley, A 15, 17, 31n10, 91
Victorian 46, 50–53, 55, 57, 59, 63, 68, 71, 77, 81, 83, 96, 103, 106, 108, 119, 135, 138, 139, 150, 154, 158, 162, 163, 175, 177, 178, 180, 183–189, 204, 211, 213
visualise 54, 57, 68, 70, 85

Walkerdine, V 222n6
walls 4, 11, 52, 54, 56, 60, 65, 76, 92, 94, 107, 112, 121, 149–151, 159, 161, 172, 177, 180, 188, 193, 212
Waterton, E 1, 3, 15, 22, 218, 222n6
Watson, S 4, 23, 86
Wetherell, M (with Smith, L and Campbell, G) 8, 28n1
windows 12, 52, 76, 94, 96, 108, 109, 118–120, 126, 131, 162, 176, 178, 181, 189, 204
women 6, 8, 12, 58, 71, 72, 74, 84, 85, 96–98, 108, 112, 141, 196n3
Woodham, A (with King, L, Gloyn, L, Crewe, V and Blair, F) 10, 198
working class 103, 209, 212, 215
Wright, P 18

Yarrow, T 6, 7, 19, 24, 48, 172

Printed in the United States
By Bookmasters